MW00717363

AEROBIOLOGY

Edited by
MICHAEL MUILENBERG
HARRIET BURGE

LEWIS PUBLISHERS

Boca Raton　New York　London　Tokyo

Acquiring Editor: Joel Stein
Project Editor: Debbie Didier
Marketing Manager: Greg Daurelle
Direct Marketing: Arline Massey
Cover Design: Dawn Boyd
Manufacturing Assistant: Sheri Schwartz

Library of Congress Cataloging-in-Publication Data

Pan-American Aerobiology Association.
    Aerobiology : proceedings of the Pan-American Aerobiology
Association / edited by Michael Muilenberg, Harriet Burge.
        p.   cm.
    Includes bibliographical references and index.
    ISBN 1-56670-206-2 (alk. paper)
    1. Air--Microbiology--Congresses.   I. Muilenberg, Michael.
II. Burge, Harriet. III. Title.
QR101.P36 1996   95-50292
576'.190961--dc20   CIP

© 1996 by CRC Press, Inc.
Lewis Publishers is an imprint of CRC Press

No claim to original U.S. Government works
International Standard Book Number 1-56670-206-2
Library of Congress Card Number 95-50292
Printed in the United States of America 1 2 3 4 5 6 7 8 9 0
Printed on acid-free paper

INTRODUCTION

Aerobiology has been defined as "the scientific discipline focused on the transport of organisms and biologically significant materials through the atmosphere" (British Aerobiology Federation statutes).[1] A somewhat expanded definition by C.S. Cox is "the science of the aerial transport of microorganisms and other microscopic biological materials together with their transfer to the air, their deposition and the ensuing consequence for life forms including the microscopic entities themselves."[2]

These definitions allow the inclusion of a large number of specialties under the "umbrella" of aerobiology. The more frequently mentioned include aerosol physics, allergy, bacteriology, biometeorology, botany, entomology, epidemiology, indoor air quality, mycology, palynology, plant pathology, respiratory physiology, and others. In addition, the study of airborne particles (of biological origin or not) to which organisms, especially humans, are exposed via the respiratory pathway either intentionally (aerosol drug delivery) or not (asbestos fibers or noxious chemicals) is sometimes considered as a part of aerobiology. Most practitioners in these fields do not consider themselves "aerobiologists" but, rather, use aerobiological techniques as part of their research. The Pan-American Aerobiology Association (as well as other regional organizations most of which are united within the International Association of Aerobiology) serves to bring together from multiple disciplines those who study airborne biological particles, promoting the exchange of ideas about aerobiology across interdisciplinary lines.

The Pan-American Aerobiology Association (PAAA) was an outgrowth of the Canadian Aerobiological Conference "Aerobiology, Health, Environment" organized by Paul Comtois at the University of Montreal (1–3 June 1988, Montreal, Quebec). Some of those attending this meeting had been a part of the series of Gordon Research Conferences on aerobiology, organized by Dr. William S. Benninghoff, and were also members of the International Association for Aerobiology (IAA), organized in 1974 in The Hague, Netherlands, which meets at 4-year intervals. A small group from the Montreal meeting recognized the benefit of meeting on a more frequent basis to exchange ideas, discuss research projects, consider cooperative studies, and encourage young scientists to become involved in aerobiological research. This group developed the initial plan for the PAAA, which was formally organized in 1989, and recognized as an affiliate by the IAA in 1990. Annual meetings have recently been held in Ann Arbor, Michigan (1991), Scarborough, Ontario (1992), and Mexico City (1993).

This volume represents a snapshot of current interests within the aerobiology community. The chapters that follow have been separated into two (overlapping) sections: those dealing primarily with ecology and distribution, and those that focus on human health. The lack of a distinct line of demarcation between these groups will soon become obvious to the reader. The keynote address at the symposium was presented by Dr. John Lacey of the Rothamsted Experimental Station (Harpenden, England) and is the first chapter in this volume. Dr. Lacey describes composting, samplers commonly used in composting environments, and the difficulties encountered when interpreting recoveries. Chapter 3 by Emberlin and Norris-Hill is a good example of the work that is being done by a number of groups on correlating aeroallergen concentrations with one or more weather parameters. Eventually, these studies could lead to better aeroallergen concentration predictions and allergy management. The Hurtado and Alson chapter (Chapter 4) on Mimosoideae (a subfamily of the pea family, Leguminosae) pollen in Venezuela is an example of studies that need to be done in areas where outdoor bioaerosols have not been adequately characterized. Comtois and Boucher (Chapter 2) examine factors that affect pollen production and "pollen seasons" using a novel approach that includes examining pollen production and release from different parts of the ragweed plant.

The characterization of the fungal component of dust from school rooms in Spain is the focus of the paper by Angulo et al. (Chapter 5), and reflects current widespread interest in dust as a source of allergen exposure. Much work has been done on house dust mites (see O'Rourke et al.; Chapter 6) but comparatively little on the fungi in dust, and less on the fungi and other antigens in schools. The connection between concentrations in dust and respiratory exposure remains unclear. The next two chapters (7 and 8) focus on exposures related to the handling of waste water and domestic solid waste (Lavoie and Rosas et al., respectively). Although both chapters deal primarily with sampling, the connection to health risk assessment is obvious. Both address the questions of worker and community exposure to potentially harmful agents, a necessary component of risk assessment. Miller and Kenepp (Chapter 9), after analyzing close to 800 cooling towers for *Legionella* and other bacteria, present recoveries as ratios of *Legionella* to total bacteria and discuss the use of such a ratio to characterize the microbial status of cooling towers. Ultimately, such ratios could contribute to a better understanding of health risks associated with the presence of *Legionella* in cooling towers.

The nature of allergens and their sources have been of primary interest for many members of the aerobiology community. Pettyjohn and Levetin (Chapter 10) have explored the allergenicity of pine pollen, the subject of a long term debate within the allergy and aerobiology communities. Vijay et al. (Chapter 11) present their work on the characterization of variability in allergen content of the fungus *Cladosporium herbarum*. Little is known about the variation in allergen content between batches, isolates, strains, species, or even genera of many fungi. Finally, Dr. Lacey's chapter (Chapter 12) on bioaerosols, both microbial and nonmicrobial, associated with occupational asthma, fittingly focuses on two especially important aspects of aerobiological studies, sampling and analytical methods.

Unlike many other symposium "proceedings," this book is not simply a compilation of unedited manuscripts resulting from meeting presentations.

Each chapter in this volume has undergone a thorough review process consistent with the retention of new and possibly controversial ideas. Where disagreements on theory or interpretation have occurred between the authors and the editors, the authors were, in most cases, allowed to prevail. Therefore, the ideas presented here are not necessarily in agreement with those of the editors or of the PAAA as a whole.

Michael L. Muilenberg, M.S.
Research Associate and Instructor

Harriet A. Burge, Ph.D.
Associate Professor of Environmental
   Microbiology

Editors
Department of Environmental Health
Harvard University School of Public
   Health
Boston, Massachusetts

REFERENCES

1. Hirst, J.M., Statutes of the British Aerobiology Federation, Aerobiology at Rothamsted, *Grana*, 33, 66, 1994.
2. Cox, C.S., *The Aerobiological Pathway of Microorganisms*, John Wiley & Sons, Chichester, 1987, xi.

ACKNOWLEDGMENTS

The success of the Pan-American Aerobiology Association symposium held at the University of Michigan (from which this volume was an outgrowth) was made possible through the efforts of Thersa Sweet, currently a doctoral student in Public Health at the University of Michigan. Her assistance in organizing the symposium was invaluable. We would also like to thank Dr. Paul Comtois, Professor of Geology at the University of Montreal and 1991 president of PAAA for his friendship and support. Financial support for the symposium by Costar Corporation, Cambridge, MA, and Graseby Andersen, Atlanta, GA, and support from the Center for Indoor Air Research during the long editing process, are also gratefully acknowledged.

A great deal of credit and thanks is due to the following reviewers (listed alphabetically) for their most helpful assistance:

- Martin D. Chapman, Ph.D., Internal Medicine, School of Medicine, University of Virginia, Charlottesville, VA
- Kirby F. Fannin, Ph.D., M.P.H., Life's Resources, Inc., Addison, MI
- Elliot Horner, Ph.D., Tulane Medical Center, Tulane University, New Orleans, LA
- Gerald J. Keeler, Ph.D., Environmental and Industrial Health, School of Public Health, University of Michigan, Ann Arbor, MI
- Mary Kay O'Rourke, Ph.D., Arizona State University, Tuscon, AZ
- Brian G. Shelton, Pathogen Control Associates, Inc., Norcross, GA
- Michael D. Walters, Ph.D., Environmental Health, Harvard School of Public Health and Polaroid Corp., Cambridge, MA

CONTRIBUTORS

Julio Alson
Laboratoria de Alergia Experimental, Centro de Microbiología y Biología Celular, Instituto Venezolano de Investigaciones, Científicas, Caracas, Venezuela

Julia Angulo-Romero
Departamento de Biología Vegetal y Ecología, Universidad de Córdoba, Spain

Larry G. Arlian
Department of Biological Sciences, School of Medicine, Wright State University, Dayton, Ohio

A. Janaki Bai
Botany Department, Andhra University, Waltair, India

Stéphane Boucher
Laboratorie d'aérobiologie, Département de géographie, Université de Montréal, Canada

Maureen Burton
Bureau of Drug Research, Sir FG Banting Research Centre, Tunney's Pasture, Ottawa, Ontario, Canada

Carmen Calderón
Centro de Ciencias de la Atmosfera, Circuito Exterior, Ciudad Universitaria, Mexico City, Mexico

José M. Caridad-Ocerin
Departamento de Biología Vegetal y Ecología, Universidad de Córdoba, Spain

Paul Comtois
Laboratorie d'aérobiologie, Département de géographie, Université de Montréal, Canada

Michael P. Corlett
Mycology/Center for Land and Biology, Ottawa, Ontario, Canada

Brian Crook
Health and Safety Executive, Health and Safety Laboratory, Sheffield, United Kingdom

Eugenio Domínguez-Vilches
Departamento de Biología Vegetal y Ecología, Universidad de Córdoba, Spain

Jean C. Emberlin
Pollen Research Unit, Worcester College of Higher Education, Worcester, England

Inés Hurtado
Laboratoria de Alergia Experimental, Centro de Microbiología y Biología Celular, Instituto Venezolano de Investigaciones, Científicas, Caracas, Venezuela

Félix Infante-Garcia-Panteleon
Departamento de Biología Vegetal y Ecología, Universidad de Córdoba, Spain

Kathy A. Kenepp
Department of Microbiology and Immunology, School of Medicine, University of Louisville, Louisville, Kentucky

John Lacey
IACR-Rothamsted, Harpenden, Hertfordshire, England

Jacques Lavoie
Institut de Recherche en Santé, et en Sécurité du Travail du Québec, Montréal, Québec, Canada

Estelle Levetin
Faculty of Biological Sciences, University of Tulsa, Tulsa, Okalahoma

Genevieve Marchand
Institut de Recherche en Santé, et en Sécurité du Travail du Québec, Montréal, Québec, Canada

Ana Mediavilla-Molina
Departamento de Biología Vegetal y Ecología, Universidad de Córdoba, Spain

Richard D. Miller
Department of Microbiology and Immunology, School of Medicine, University of Louisville, Louisville, Kentucky

Cathy L. Moore
Department of Geosciences, University of Arizona, Tucson, Arizona

Gauri Muradia
Bureau of Drug Research, Sir FG Banting Research Centre, Tunney's Pasture, Ottawa, Ontario, Canada

Jane Norris-Hill
St. David's College, Lampeter, Wales, United Kingdom

Mary Kay O'Rourke
Respiratory Sciences Center, AHSC, University of Arizona, Tucson, Arizona

Mary E. Pettyjohn
Faculty of Biological Sciences, University of Tulsa, Tulsa, Oklahoma

Sophie Pineau
Hydro-Quebec, Quebec, Canada

Irma Rosas
Centro de Ciencias de la Atmosfera, Circuito Exterior, Ciudad Universitaria, Mexico City, Mexico

E. Salinas
Centro de Ciencias de la Atmosfera, Circuito Exterior, Ciudad Universitaria, Mexico City, Mexico

Hari M. Vijay
Bureau of Drug Research, Sir FG Banting Research Centre, Tunney's Pasture, Ottawa, Ontario, Canada

Pauline A. M. Williamson
IACR-Rothamsted, Harpenden, Hertfordshire, England

Martin Young
Institue for Biological Science, Ottawa, Ontario, Canada

CONTENTS

AEROBIOLOGY

Chapter 1

MICROBIAL EMISSIONS FROM COMPOSTS AND ASSOCIATED RISKS – TRIALS AND TRIBULATIONS OF AN OCCUPATIONAL AEROBIOLOGIST

John Lacey
Pauline A.M. Williamson
Brian Crook

CONTENTS

1-56670-206-2/96/$0.00+$.50
© 1996 by CRC Press, Inc.

1

I. ABSTRACT

Composting is used to produce substrates suitable for mushroom cultivation or to aid in the disposal of domestic waste. Handling such composts can cause the release of microorganisms in large numbers and present hazards to workers, especially of hypersensitivity pneumonitis and aspergillosis. On mushroom farms, up to 10^8 actinomycete spores (mostly *Thermomonospora* spp.) and bacteria/m^3 air can be released when compost is disturbed, especially during spawning. During our studies, *Saccharopolyspora* (*Faenia*) *rectivirgula* and *Thermoactinomyces* spp., sometimes implicated in mushroom worker's lung, were usually few. *Talaromyces, Scytalidium* spp., and *Aspergillus fumigatus* were the predominant fungi, each numbering up to $6.5 \times 10^3/m^3$ air. Personal samplers showed that workers were heavily exposed during spawning. Up to 10^8 actinomycetes and bacteria/m^3 were also released into the air when composted domestic waste was handled. The most abundant species differed with composting conditions but *Saccharomonospora, Thermoactinomyces, Thermomonospora, Saccharopolyspora,* and *Streptomyces* were all abundant in some samples. *A. fumigatus* and *Penicillium* spp. tended to be more numerous in composted domestic waste than in mushroom composts (up to 2.7×10^6 spores/m^3). The study of occupational lung disease among workers handling composts is complicated by many problems, including overloading of samplers due to the high spore concentrations, intermittent sources, identification of organisms that might be implicated, poor growth and antigen production in culture, disease diagnosis, and risk assessment. These problems are discussed and suggestions made for future study.

II. INTRODUCTION

Composting provides an example of occupational exposure to bacteria and actinomycete and fungal spores that illustrates many of the problems that beset the occupational aerobiologist. Composting is an exothermic process in which organic substrates are subjected to microbial degradation. Microbial activity leads to the production of heat, with temperatures up to 70°C or more, through the energy released by respiration. Mushroom and domestic waste composts differ in their substrates and in the control of temperature. Microorganisms are released into the air when compost piles are formed or dismantled, when compost is moved, especially where it falls from one conveyor to another, when mushroom spawn is added and during pre- and post-compost processing of domestic waste. The nature of the airborne microflora depends on the existing contamination of the starting materials, microbial development between disposal and composting and, subsequently, development during composting. Allergen-producing organisms, especially when present in large numbers, can present a hazard to the health of exposed workers, through occupational asthma or extrinsic allergic alveolitis (hypersensitivity pneumonitis).[1] Infection rarely occurs.

This chapter describes approaches to the problems of sampling large concentrations of spores, identification of microorganisms and their antigens, disease diagnosis and risk assessment, and makes suggestions for further work.

III. MATERIALS AND METHODS

The samplers we have used at mushroom and domestic waste composting sites (although not all discussed in detail in this chapter) have included cascade impactors, collecting on non-nutrient surfaces (20 l/min, up to 2 min),[2] Andersen culture plate cascade impactors (Graseby Andersen, Smyrna, GA) (25 l/min, 5–60 s),[3] multi-stage liquid impingers (55 l/min, up to 30 min),[4] Millipore® (Millipore Corp., Bedford, MA) or Nuclepore® (Corning Costar Corp., Cambridge, MA) personal filtration aerosol monitors (with 0.8 μm diameter pores; 2 l/min, up to 3 h),[5] high-volume filtration air samplers (650 l/min, up to 30 min), large-volume electrostatic sampler (Sci-Med Environmental Systems, Eden Prairie, MN) (600 l/min, up to 30 min) and a Royco® particle counter (Pacific Scientific, Marlow, U.K.) (0.28 or 2.8 l min for 30 or 60 s every 2 or 5 min). Isolation media and temperatures used have included:

- For fungi: (1) 2% malt extract (Oxoid®, Unipath Ltd., Basingstoke, Hampshire, U.K.) agar + penicillin (20 IU/ml) + streptomycin (40 units/ml)[6] (25 and 37°C); (2) dichloran rose bengal chlortetracycline agar (Oxoid) (25 and 37°C)
- For total bacteria and actinomycetes: (1) nutrient agar (Oxoid®, half-strength) + actidione (100 μg/ml)[6] (25 and 40°C); (2) tryptone soya agar (Oxoid, half-strength) + actidione (100 μg/ml) (37°C); (3) tryptone soya agar (Oxoid, half-strength) + 0.4% casein hydrolysate agar (TSC) + actidione (100 μg/ml)[6] (55°C)
- For Gram-negative bacteria: violet red bile glucose agar (Oxoid) (37°C)
- For streptococci: KF Streptococcal agar (Oxoid) (37°C)
- For *Salmonella*: Sample from multi-stage liquid impinger incubated in Rappaport's medium at 43°C for 8 h then plated on modified xylose lysine (XLD) and Brilliant green agars
- For *Thermomonospora* spp.: TSC + rifampicin (5 μg/l) + actidione (100 μg/ml) (55°C)[7]
- For *S. rectivirgula*: modified hippurate agar (NaCl, 20.0 g; MgSO₄.7H₂O, 0.2 g; NH₄H₂PO₄, 1.0 g; sodium hippurate (hippuric acid,

Na salt), 3.0 g; K_2HPO_4, 1.0 g; agar, 24.0 g; phenol red, 5.0 ml; distilled
water, 1000 ml) + actidione (100 µg/ml) (HAX) (55°C)
- For *Thermoactinomyces* spp: Czapek yeast casein agar (Czapek agar,
  Oxoid, 45.4 g; yeast extract, 5 g; casein hydrolysate, 2 g; water, 1000
  ml) + tyrosine (0.3%) + novobiocin (25 µg/ml) (55°C)[8]

Samples using Andersen samplers together with each medium/incubation
temperature combination and, where possible, with personal filtration samplers
were collected in triplicate at each site and stated concentrations are the means
of these determinations. Unless stated otherwise, spore/cell concentrations are
from Andersen sampler plates, corrected for multiple deposition of particles
at deposition sites using the table from Andersen,[3] or from personal filtration
samplers, after resuspension, dilution, and plating following the methods of
Palmgren et al.[5]

IV. MUSHROOM COMPOSTS

Mushroom compost is usually made from horse manure, wetted straw,
gypsum, and a nitrogenous supplement, often poultry manure, which is stacked
for 7–10 days with regular turning in long piles that heat to 75–80°C (Phase
1). In Phase 2, the partially composted material is placed in large chambers
(tunnels) and is heated to about 60°C with hot, humidified air for 3–10 days.
Some thermophilic actinomycetes grow during Phase 1 but most develop
during Phase 2 to form a visible white weft on the compost.[9] After cooling,
the compost is mixed with spawn and placed in growing houses in trays or
on continuous shelves. The fruiting bodies are picked by hand (cropping).
After cropping is complete, the compost is pasteurized with steam (cookout)
and removed from the trays or shelves for disposal. To evaluate worker exposure, general environmental samples were collected using Andersen and other
samplers in each area (only Andersen sampler results are presented here), and
personal filtration aerosol monitors were fitted on representative workers and
operated for periods up to 3 h, depending on shift and break patterns. To
simplify interpretation of exposure measurements, workers were placed in
different job categories depending on the nature of their exposure to compost.
Job category A included workers who prepared the compost and added the
mushroom spawn; Job category B included workers who only handled the
compost after cropping was complete and were responsible for pasteurization
of the compost and emptying boxes; Job category C, the largest category,
included all who picked mushrooms.

A. JOB CATEGORY A: COMPOST PREPARATION
Straw bales for composting often released many spores when opened
(straw blending; Figure 1). On one farm, bales were fed into a machine with
a control panel at the side which broke them open and wetted them. Large
numbers of spores were released near the control panel as each bale was

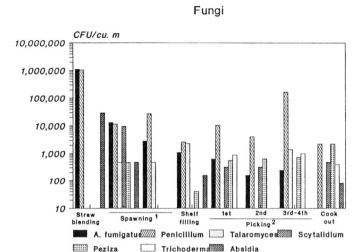

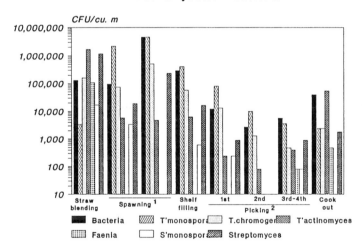

FIGURE 1 Airborne microorganisms released during preparation and use of mushroom compost. *Faenia = Saccharopolyspora.* 1, Samples taken in two separate spawning areas on Farm H. 2, Samples taken in growing houses during, respectively, the first, second, and third to fourth flushes.

opened but their numbers and types depended on the quality of the straw. Concentrations of *A. fumigatus* in this area exceeded 10^6 colony forming units (cfu)/m^3 and there were similar numbers of *Penicillium, Thermoactinomyces,* and *Streptomyces* spp. *S. rectivirgula, Thermomonospora* spp., *Aspergillus terreus,* and a range of other fungi were also present (Andersen samples). Personal exposure of the operator reflected this microflora with minor differences probably resulting from exposure to more bales during the longer period of sampling. When bales were broken open with a mechanical shovel, workers were well isolated from spores unless they were directly downwind. These differences may also result from the different sampling methods; personal-filteration monitors allowing accurate detection of more concentrated aerosols, but possibly damaging sensitive organisms.

B. JOB CATEGORY A: SPAWNING

Between the undisturbed compost windrows, bacteria numbered about 5.4 $\times 10^3$ cfu/m^3 air and actinomycetes, mostly white *Thermomonospora* spp., 6.6 $\times 10^4$ cfu/m^3. However, after Phase 2, when spawn was mixed with the compost, numbers of actinomycete spores and cfu in the air increased dramatically. Microscopic assessments of bacterial cell and actinomycete and fungal spore numbers on cascade impactor slides collected on two farms revealed 2.2 and 4.3×10^8 bacteria and actinomycete spores/m^3 air, respectively. However, fungal recoveries never exceeded 1.4×10^5 cfu/m^3 with individual taxa numbering 10^3–10^5 cfu/m^3. *Aspergillus fumigatus* and *Penicillium* spp. were among the most numerous, but *Scytalidium, Talaromyces, Absidia,* and yeasts were also isolated frequently (Figure 1). Culturable bacteria and actinomycetes usually numbered 10^6–10^7 cfu/m^3 air, with *Bacillus* spp., white *Thermomonospora* spp., *T. chromogena,* and *Actinomadura*-like isolates most numerous (Figure 1). No *S. rectivirgula* were isolated at any spawning site and few *Thermoactinomyces* and *Saccharomonospora* spp. (Figure 1). Personal samplers confirmed that workers in the spawning sheds were heavily exposed to spores and sometimes to greater numbers than indicated by general area samples with Andersen samplers. Personal samplers yielded up to 6×10^7 cfu actinomycetes/m^3 and 3.2×10^7 other bacteria but few fungi. On one farm the personal sampler recovered up to 32×10^7 cfu total bacteria, 3.4×10^7 cfu *Thermomonospora* and 1.7×10^7 cfu *Actinomadura*/m^3 air. *Penicillium* spp. and *A. fumigatus* counts were up to 8.9 and 3.8×10^4 cfu/m^3 air, respectively.

C. JOB CATEGORY B: COOKOUT

During cookout, bacteria and actinomycetes on one farm numbered about 6.9×10^5 cfu/m^3 air, and included *Bacillus* (2.6×10^4/m^3), *Thermomonospora* (1.8×10^5/m^3), *Actinomadura* (2.3×10^4/m^3), and *Saccharomonospora* (5.2×10^4/m^3). *Monilia* and *Cladosporium* were the most numerous fungi in personal samples, both accounting for about 1.2×10^5 cfu/m^3. *Epicoccum* and *Penicillium* contributed a further 5.9 and 3.8×10^4 cfu/m^3, respectively, but *A. fumigatus* only about 9×10^3 cfu/m^3. By contrast, on another farm, numbers of bacteria and actinomycetes had decreased by cookout to about 10^4 cfu/m^3 air, with *Thermoactinomyces* spp. most numerous; the predominant fungi included *Penicillium* spp., *Peziza ostracoderma (Plicaria fulva), Humicola lanuginosa, Cladosporium,* and *Trichoderma.* The forklift driver in the cookout area was exposed to more than 10^5 cfu bacteria and actinomycetes/m^3 air and to fewer than 10^4 fungi.

D. JOB CATEGORY C: GROWING HOUSES

During cropping (picking), bacteria and actinomycetes were few (only 100 cfu/m^3) and *Penicillium* spp. were the most numerous fungi, yielding about 10^5 cfu/m^3 air, while *Plicaria* and *Trichoderma* each yielded about 10^3 cfu/m^3. Personal exposure to bacteria and actinomycetes was variable and, surprisingly, greater in houses in full production than in those earlier in the cropping cycle, perhaps because the compost had become disturbed during picking. Early in the cropping cycle, workers were exposed to 38–258 cfu *Bacillus* and 0.9–1.7 × 10^3 cfu *Thermomonospora*/m^3 and at peak cropping from 70–115 and 0.1–728 × 10^3 cfu/m^3, respectively. Of the fungi, exposure to *Trichoderma* tended to be greater early in cropping than at the peak (17–123 × 10^3 cfu/m^3 compared to 9–49 × 10^3) while that to *Penicillium* was greater at peak cropping (120–221 × 10^3 cfu/m^3 compared to 95–381 × 10^3 cfu/m^3). *A. fumigatus* exposure was also greater during the early flushes than at peak production (5–471 × 10^3 cfu/m^3 compared to 1–43 × 10^3). Disease control workers, who removed diseased mushroom fruiting bodies from all growing sheds, were exposed to smaller mean concentrations than pickers.

On another farm, most types of bacteria and actinomycetes numbered 10^2–10^4 cfu/m^3. However, one house in its first flush yielded 8 × 10^4 cfu *Thermomonospora*/m^3 and 880 cfu *Trichoderma*/m^3, both concentrations much larger than usual in growing houses. *Penicillium* spp. usually formed the most numerous group of fungi. Numbers of fungi generally increased with age of crop, with most (10^5 *Penicillium* cfu/m^3, 10^3 *Trichoderma*, 720 *Peziza*) in a house in its third or fourth flush. Personal samplers on pickers emphasized the large numbers of *Thermomonospora* and *Trichoderma* during the first flush on this farm and the general increase in numbers with age of the crop.

V. DOMESTIC WASTE COMPOST

Domestic waste is a complex mixture of organic and inorganic materials.[10] Magnetic separation and screening removes metal, glass, paper, and plastic that will not compost and the remaining putrescible material <50 mm in size is usually stacked with aeration and, sometimes, a water supply and allowed to compost for about four weeks. Temperatures are usually controlled at about 50–60°C, to achieve optimum decomposition[11–13] and to kill pathogenic bacteria rapidly[10,14] which form about 6% of the total bacteria on waste reaching transfer stations.[15] One week above 65°C is sometimes recommended[16] although this may decrease the diversity of species available for decomposition.[17] Samples were collected with Andersen samplers and personal filtration monitors during the operation of a pilot plant at Warren Spring Laboratory, Stevenage, UK, during the processing of waste before and after composting and with multi-stage liquid impingers and personal filtration monitors while dismantling a pile after composting. The multi-stage liquid impingers were fitted with hemispherical baffles to allow stagnation point sampling and were loaded with one fourth strength Ringer's solution + 2% inositol. The layout of the pilot plant and the position of sampling sites during pre- and post-composting processing of domestic waste is shown in Figure 2. Microorganisms could become airborne at many points during the processing but especially where compost passed over the vibrating conveyor and through the rotating trommel and wherever the compost fell from one conveyor to another. There were few workers in the plant and, unless there were breakdowns, workers were mostly near sample point 3 supervising loading of the trolleys or removing them with forklifts.

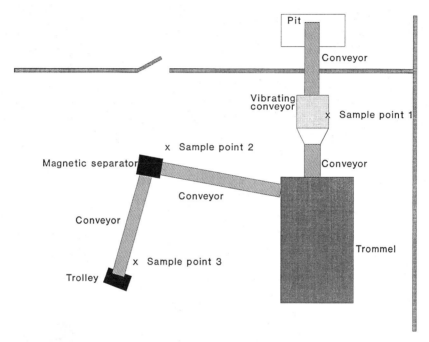

FIGURE 2 Layout of the Pilot Plant at Warren Spring Laboratory showing positions of the three sampling points (x).

A. PRECOMPOSTING PROCESSING

During precomposting processing, bacteria and actinomycetes increased from about 10^4 cfu/m^3 to between 5.7×10^4 and 2.3×10^5 cfu/m^3, thermophilic actinomycetes from about 10^4 cfu/m^3 to between 4.8×10^4 and 7.5×10^5 cfu/m^3, and fungi from about 10^3 cfu/m^3 to between 1.5×10^4 and 1.9×10^5 cfu/m^3, with *Penicillium, Cladosporium,* and yeasts showing the largest increases (Figure 3). Gram-negative bacteria accounted for more than 10^4 cfu/m^3 air and streptococci to 7.6×10^3 cfu/m^3. Dust concentrations in the breathing zones of workers consistently exceeded 10 mg/m^3 air, the recom-

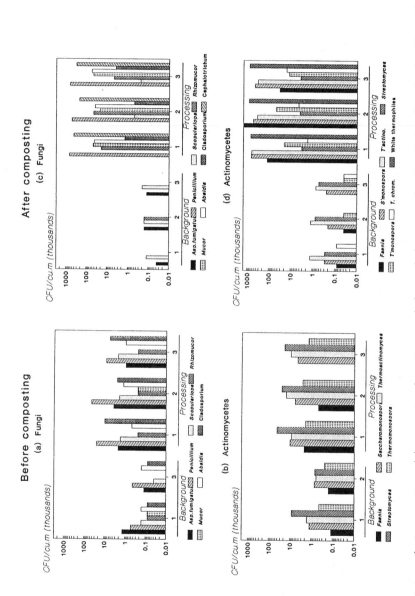

FIGURE 3 Emissions of microorganisms while processing domestic waste before and after composting, Run 2. Sampling sites (see Figure 2) are on X-axis. *Scopulariops, Scopulariopsis; S'monospora, Saccharomonospora; T'actino., Thermoactinomyces; T'monospora, Thermomonospora; T. chrom., Thermomonospora chromogena; Faenia = Saccharopolyspora.*

mended limit for nuisance dusts.[18] Personal exposure of workers near sample point 3 was particularly heavy during the first run. Personal exposure to bacteria ranged from 2.3×10^5 to 1.6×10^6 cfu/m³ air, to actinomycetes up to 5.3×10^5 cfu and to fungi up to 5.3×10^4 cfu/m³.

B. DISMANTLING THE PILE

Dismantling manually and with a tractor-mounted shovel in dry weather with a light wind resulted in the emission of many microorganisms (Table 1) with large catches both 10 m downwind and 5 m upwind. Maximum concentrations 10 m downwind exceeded 4.2×10^6 cfu microorganisms/m³ and 5 m upwind about 1.3×10^6 cfu/m³. Of these, up to 87% of the fungi were *A. fumigatus* and 70% of the thermophiles were actinomycetes, mostly *Thermoactinomyces* and *Thermomonospora* spp. Gram-negative bacteria, mostly *Pseudomonas* spp., numbered up to 8000/m³ and faecal streptococci up to 1550/m³. Personal samplers indicated that workers handling the compost were much more heavily exposed than indicated by the liquid impingers downwind, although it should be noted that liquid impingers may underestimate levels of hydrophobic particles. One worker on the pile was exposed to more than 2.7×10^7 spores/m³, mostly actinomycetes

C. POSTCOMPOSTING PROCESSING

All the major groups of organisms exceeded 10^5 cfu/m³ air and often 10^6 (Andersen samples), especially with the second batch of compost (Run 2) examined which was noticeably dustier than the earlier batch. Gram-negative bacteria and streptococci generally exceeded 10^4 cfu/m³ air, with the latter reaching 2×10^5/m³ during Run 2. Species composition changed considerably during composting (Figure 3). Actinomycetes increased vastly in numbers to exceed other bacteria. *Saccharopolyspora*, *Saccharomonospora*, *Thermoactinomyces,* and *Streptomyces* spp. all exceeded 10^5 cfu/m³ air and in some samples exceeded 10^6 cfu/m³. *Thermomonospora* spp. were less numerous, with concentrations generally about 10^4 to 10^5 cfu/m³. However, in the second run, plates were often so crowded that the white thermophilic actinomycetes, *Thermoactinomyces*, *Thermomonospora,* and *Actinomadura* spp., could not be differentiated. Therefore, the concentrations of each are probably underestimates. Fungi also differed between the two runs with *A. fumigatus* being replaced by *Penicillium* spp. in the second run, possibly a consequence of the addition of water or lower temperatures. *Cephalotrichum* spp. were perhaps underestimated on malt extract plates incubated at 25 or 37°C since Czapek-yeast-casein-tyrosine agar plates containing novobiocin (CYCT+N), first incubated at 55°C and then stored at 4°C, yielded extremely large numbers of this fungus (up to 4.2×10^5 cfu/m³ air) where malt plates had yielded only up to 2.9×10^3 cfu/m³. Spore germination was apparently stimulated by the high temperature incubation although growth occurred only after storage at the lower temperature. Personal samplers on workers yielded up to 1.3×10^7 cfu bacteria/m³ air during Run 1 and up to 4.2×10^8 during Run 2 and up to

TABLE 1   Emissions of Microorganisms from a Compost Pile During Dismantling

Spore concentration (cfu × 10⁻⁴/m³)

Sampling Position	Fungi 37°C		Total Bacteria 37°C		Gram Negative Bacteria 37°C		Thermophilic Bacteria 55°C	
	Start	Finish	Start	Finish	Start	Finish	Start	Finish
Multi-stage liquid impinger:								
5 m upwind	<0.01	4.4	0.03	31.6	<0.01	0.3	0.07	123
10 m downwind	0.2	41.0	2.7	140.1	0.2	0.8	0.2	382
Aerosol monitor:								
Next to pile	0.2	651	2.2	761	0.4	6.9	0.09	466
Worker on pile	2.0	819	5.2	1180	2.0	23.4	0.4	1880
Worker screening compost	—	124	—	355	—	7.1	—	639

Note: Start, during construction of the pile; Finish, dismantling after composting.

3.3×10^6 and 1.8×10^7 cfu fungi/m^3, respectively. Mesophilic actinomycetes numbered up to 6×10^6 cfu/m^3 in Run 1 and 2.7×10^8 in Run 2 and thermophiles up to 1.2×10^7 and 1.6×10^8 cfu/m^3 (Figure 3), respectively. Species composition was similar to that obtained with the Andersen sampler with *S. rectivirgula* numbering up to 4.6×10^6 and 3.5×10^7 cfu/m^3, in the first and second runs, respectively, *Thermoactinomyces* spp. 1.6×10^6 and 2.5×10^7, *Saccharomonospora* spp. 3.7×10^6 and 8.5×10^7, and *Thermomonospora* spp. 7.0×10^6 and 1.2×10^7. Similarly, *A. fumigatus* predominated during Run 1 with up to 2.8×10^6 cfu/m^3 and *Penicillium* during run 2, with up to 9.3×10^6 although *A. fumigatus*, with up to 7.4×10^6 cfu/m^3, was isolated in much greater numbers than expected from Andersen sampler counts.

VI. PROBLEMS OF SAMPLING OCCUPATIONAL ENVIRONMENTS

Among the problems that affect the occupational aerobiologist are intermittent and unexpected sources of microorganisms; sampling in highly contaminated atmospheres; the detection, accurate quantification, and identification of antigen sources; the preparation of antigenic extracts; disease diagnosis, and risk assessment. Some of these are interrelated. For instance, large concentrations of spores enforce short sampling times to minimize overloading, and these are subject to great variation, especially if the source is intermittent or there is turbulent wind. During bale opening on mushroom farms, spores are released in large numbers as each individual bale is opened but then in decreasing numbers and from different parts of the machine as the straw is mixed with water. Sampling at different stages of this process would result in very different estimates of spore exposure. Similarly, if wind direction is variable when sampling out of doors, estimates of spore concentration will depend on the source of the air sampled, how much aerosol it carries, and the direction of the sampler orifice with respect to wind direction.

Unexpected sources of microorganisms can sometimes be found. On one mushroom farm, the water used to wet the straw was a major source of bacterial aerosols. The excess water from the straw pile was collected in an aerated pond, where bacteria multiplied, and then recirculated to the sprays. There was a tenfold increase in airborne bacteria 20 m downwind, from 2.7×10^3 to 3.5×10^4/m^3 air during spraying.

Direct impaction samplers, such as the Andersen sampler, are very susceptible to overloading and within such highly contaminated atmospheres as those associated with compost handling, this can occur in a few seconds. Ideally, the deposition sites on an Andersen sampler plate should not be more than half occupied by colonies to minimize errors caused by multiple impaction at the same site. Other samplers that are not time limited, such as liquid impingers and filtration samplers can be used but these also have disadvantages:

1. Apart from the multi-stage liquid impinger, these are often not size selective so that the lower respiratory hazard cannot be quantified;
2. Impingement of particles and the preparation of spore suspensions from filters after sampling disrupt particles to give cell counts rather than particle counts preventing comparison with the Andersen sampler;
3. Both impingers and filters are less satisfactory than the Andersen sampler for isolating some actinomycetes, as well as some bacteria and fungi — possibly due to pass-through (impingers) or desiccation (filters).

However, impingers and filtration samplers also have advantages:

1. They can operate over longer periods and provide a time-weighted average, evening out fluctuations in the source strength;
2. Isolations on different selective media can be made from the same rather than from successive samples as with the Andersen sampler.

Identification of the organisms present in occupational environments has often been problematical, especially with actinomycetes. Actinomycete taxonomy has been in a state of flux for much of the past 30 years and frequently taxa encountered in stored products have been new to science or have been regarded as infrequent. Thus the organism implicated as the chief source of farmer's lung antigen has been known successively as *Thermopolyspora polyspora, Micropolyspora faeni, Faenia rectivirgula,* and now as *Saccharopolyspora rectivirgula* through misidentification, poor descriptions of new taxa, and changing concepts of taxonomic relationships. Similarly, the name, *Thermoactinomyces vulgaris* has been used in three senses: as a single variable species and for two segregates of that species. It should correctly be used for isolates which do not produce melanin pigments from tyrosine and which cannot utilize starch, synonymously with *T. candidus*. Melanin and amylase positive cultures should be referred to *T. thalpophilus* or other species.[19] The role of *Thermoactinomyces* spp. in occupational lung disease still requires reevaluation. Other species may be difficult to distinguish without chemical and physiological tests because they are similar in morphology.

Identification of the cause of mushroom worker's lung has been made difficult because the predominant *Thermomonospora* species do not grow well in culture and do not produce strong antigens for immunological testing. *S. rectivirgula* and *Thermoactinomyces* spp. have been implicated in mushroom worker's lung, but are rarely abundant in emissions from mushroom composts, and then chiefly associated with nonrespirable particles. However, the numbers of airborne microorganisms, especially *Thermomonospora* spp., released during the preparation and handling of composts are similar to those in other environments where allergic alveolitis occurs,[1] with individual workers often exposed to much larger concentrations. Most such environments have $>10^6$ spores/m^3 air.[1] However, up to 10^8 spores/m^3 may be necessary for sensitization.[20]

The form of allergic alveolitis known as mushroom worker's lung is not the only form of lung disease to occur on mushroom farms. Allergic alveolitis characteristically occurs in the spawning sheds while occupational asthma is associated with picking. Tests for specific IgE antibodies are therefore necessary as well as immunodiffusion or ELISA tests for IgG antibodies. Evidence of antibodies in sera of exposed subjects against specific microorganisms can be useful in diagnosing the cause of alveolitis in that it is evidence of exposure to the organism but its relevance to disease needs to be confirmed by other tests. Without good antigens, this is difficult.

A. fumigatus is known to cause infection but the risk of this happening has never been quantified. Fever can be caused by inhalation of endotoxins (cell wall lipopolysaccharides from Gram-negative bacteria). Airborne Gram-negative bacteria consistently exceeded a recommended safe level of $1000/m^3$ air,[18] especially after composting domestic waste. Also, dust concentrations in the breathing zones of workers handling such waste consistently exceeded 10 mg/m^3 air, the recommended limit for nuisance dusts, reaching 80 mg/m^3 in one sample. Mycotoxins may occur in fungal spores but there is little published evidence of their effects when inhaled by man. The uncertainty in the dose–response relationships and host factors in these diseases make risk assessment difficult and few estimates have been made. Similar risks are probably associated with domestic waste composting, especially with the greater incidence of *S. rectivirgula*.

VII. SUGGESTIONS FOR FUTURE STUDY

In the future, attention needs to be paid to the development of samplers capable of dealing with large concentrations of airborne spores. Currently, assessment of spore concentrations in occupational environments requires skilled operators and is time consuming, whether the assessment is based on microscope counts of spores or counts of colonies growing in Petri dishes from sampled air. New assays and molecular diagnostic procedures, perhaps immunoassays utilizing monoclonal antibodies would decrease this requirement for time and skill. New methods of characterizing dusts also need consideration, perhaps involving assay of antigens, endotoxin, glucan, enzymes, and other proteins and mycotoxins.

ACKNOWLEDGMENTS

We are grateful to the Health and Safety Executive and the Department of the Environment for funding, to Warren Spring Laboratory and the various mushroom farms for facilities to do the work, to all who readily assisted with the personal sampling, and especially to Mrs. Shelagh Nabb for technical assistance.

REFERENCES

1. Lacey, J. and Crook, B., Fungal and actinomycete spores as pollutants of the workplace and occupational allergens, *Ann. Occup. Hyg.*, 32, 515, 1988.
2. May, K.R., The cascade impactor: an instrument for sampling coarse aerosols, *J. Sci. Instrum.*, 22, 187, 1945.
3. Andersen, A.A., New sampler for the collection, sizing and enumeration of viable airborne particles, *J. Bacteriol.*, 76, 471, 1958.
4. May, K.R., Multistage liquid impinger, *Bact. Rev.*, 30, 559, 1966.
5. Palmgren, U., Ström, G., Blomquist, G., and Malmberg, P., Collection of airborne microorganisms on Nuclepore filters, estimation and analysis — CAMNEA method, *J. Appl. Bact.*, 61, 401, 1986.
6. Lacey, J. and Dutkiewicz, J., Methods for examining the microflora of mouldy hay, *J. Appl. Bacteriol.*, 41, 13, 1976.
7. Athalye, M., Lacey, J., and Goodfellow, M., Selective isolation and enumeration of actinomycetes using rifampicin, *J. Appl. Bacteriol.*, 51, 289, 1981.
8. Cross, T. and Attwell, R.W., Recovery of viable thermoactinomycete endospores from deep mud cores, in *Spore Research 1973*, Barker, A.N., Gould, G.W. and Wolf, J., Eds., Academic Press, London, 1974, 11.
9. Lacey, J., Actinomycetes in soils, composts and fodders, in *Actinomycetales: Characteristics and Practical Importance*, Skinner, F.A. and Sykes, G., Eds., Society of Applied Bacteriology Symposium Series No. 2, Academic Press, London, 1973, 231.
10. Bardos, R.P. and Lopez-Real, J.M., The composting process: susceptible feedstocks, temperature, microbiology, sanitisation and decomposition, in *Compost Processes in Waste Management*, Bidlingmaier, W. and L'Hermite, P. Eds., Commission of the European Communities, Brussels, 1989, 179.
11. Kane, B.E. and Mullins, J.T., Thermophilic fungi in a municipal waste compost system, *Mycologia*, 65, 1087, 1973.
12. Strom, P.F., Identification of thermophilic bacteria in solid-waste composting, *Appl. Environ. Microbiol.*, 50, 905, 1985.
13. Stutzenberger, F.T., Cellulolytic activity in municipal solid waste composting, *Can. J. Microbiol.*, 16, 553, 1970.
14. Biddlestone, A.J. and Gray, K.R., Composting, in *Comprehensive Biotechnology*, Vol. 4, Moo-Young, M., Ed., Pergamon, Oxford, 1986, 1059.
15. Crook, B., Higgins, S., and Lacey, J., Airborne gram negative bacteria associated with the handling of domestic waste, in *Advances in Aerobiology*, Leuschner, R., Ed., Birkhäuser, Basel, 1987, 371.
16. De Bertoldi, M., Ferranti, M.P., L'Hermite, P., and Zucconi, F., Eds., *Compost: Production, Quality and Use*, Elsevier Applied Science, London, 1987.
17. Strom, P.F., Effect of temperature on bacterial species diversity in thermophilic solid-waste composting, *Appl. Environ. Microbiol.*, 50, 899, 1985.
18. Rylander, R., Lundholm, M., and Clark, C.S., Exposure to aerosols of micro-organisms and toxins during handling of sewage sludge, in *Biological Health Risk of Sludge Disposal to Land in Cold Climates*, Wallis, P.M. and Lohmann, D.L., Eds., University of Calgary Press, Calgary, 1983, 69.
19. Lacey, J. and Cross, T., Genus *Thermoactinomyces* Tsiklinsky 1899, 501[AL], in *Bergey's Manual of Systematic Bacteriology*, Vol. 4, Williams, S.T., Sharpe M.E. and Holt J., Eds., Williams & Wilkins, Baltimore, 1989, 2574.
20. Rylander, R., Lung diseases caused by organic dusts in the farm environment, *Am. J. Ind. Med.*, 10, 221, 1986.

Chapter **2**

PHENOLOGY AND AEROBIOLOGY OF SHORT RAGWEED (*AMBROSIA ARTEMISIIFOLIA*) POLLEN

Paul Comtois
Stéphane Boucher

CONTENTS

1-56670-206-2/96/$0.00+$.50
© 1996 by CRC Press, Inc.

I. ABSTRACT

Flowering phenology and pollen dispersal are linked through the aerobiological pathway. Therefore, airborne pollen data are often interpreted as a set of phenological data. However, the precise relationship between emission and dispersal from specific plants has rarely been evaluated. In this chapter it is shown that different hierarchical levels of inflorescences on individual ragweed plants contribute differentially to the seasonal pollen curves: level 1 (apical) and level 2 flowers (linked to the stem) release pollen at the onset of the airborne pollen season, while pollen from level 3 flowers (linked to branches) is mainly found at the peak of the season. After the seasonal peak, climatic parameters apparently take predominance over physiological or genetic factors, and no relationships can be seen between airborne pollen recoveries and flowering phenology.

II. INTRODUCTION

The aerobiological pathway deals with the transport of airborne biological particles between sources and points of impaction. It is generally assumed that day to day variations in levels and kinds of airborne pollen grains represent the sequential production and release from local sources; namely trees, shrubs, or herbs. However, this assumption has never really been tested.

If the relationship between pollen production and release, and airborne levels is sufficiently powerful, only aeropalynological data would be needed to follow the flowering phenology of anemophilous plants. Also, a strong relationship would probably mean that common climatic factors govern both pollen emission and dispersal.

The real test of our hypothesis will be at the emission/dispersal interface, where plant physiology will be affected by the predominant climatic influences. The physiological status of a flowering plant is difficult to assess, but pollen emission, as estimated by open inflorescences with dehiscent stamens ready to disperse their pollen, is quite easy to determine.

Ragweed is the most abundant airborne pollen in many parts of eastern North America and comprises at least 30% of the total annual pollen production in Montréal.[1] Ragweed pollen contains potent allergens, and its flowering phenology is easy to follow. In addition, it was shown that autocorrelated cycles appear from year to year in the ragweed pollen airborne prevalence curves.[2,3] Could this be related to recurrent pollen emission phases? The dispersal of the pollen of this species has been well studied in the past.[4-6] However these studies were done in order to predict concentrations away from a source, and the phenology of the ragweed plant was not followed. For these reasons, short ragweed (*Ambrosia artemisiifolia*) was chosen to test the hypothesis that there is a relationship between airborne pollen cycles and the hierarchical position of pollen-emitting flowers on the plants.

III. MATERIALS AND METHODS

Sites were selected to facilitate observance of a large number of plants in their normal habitat and also be within areas that could be represented by our roof-top volumetric pollen sampler (250 km²).[1] In 1990, the ragweed population studied was located near a Canadian National Railway track near the Université de Montréal campus (Outremont). Because this population was eradicated in 1991 (there has been an active campaign against ragweed in Montréal since the 1970s), a second population, on a vacant lot near the Rivière-des-prairies (Pierrefonds) was studied that year (Figure 1).

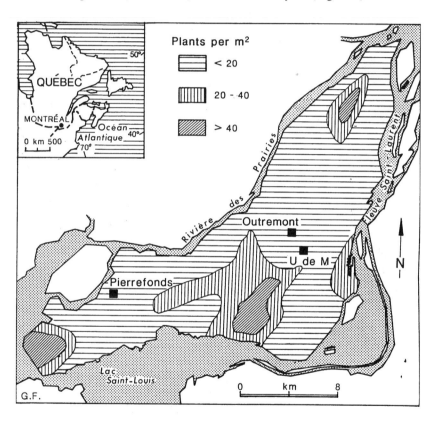

FIGURE 1 Site locations and ragweed mean densities on Montréal Island. Outremont was used in 1990, Pierrefonds in 1991. In both years, the pollen sampler was located at Université de Montréal.

In both years, phenological observations were done visually. The percentage of open stamens (or flowering status) on each inflorescence of 30 randomly selected ragweed plants was classified according to the following table:

Category 1	0% stamens emitting pollen
Category 2	1–10% stamens emitting pollen
Category 3	11–30% stamens emitting pollen
Category 4	31–50% stamens emitting pollen
Category 5	>50% stamens emitting pollen

In addition, each inflorescence was assigned to a hierarchical level according to its relative position on the plant. Level 1 corresponded to the apical inflorescence of the stem, level 2 to the apical inflorescence on a branch of first order (attached to the main stem), and level 3 to inflorescences born on a lateral branch, not directly linked to the stem (branch of second order) (Figure 2). Data on flowering status (that is category) were recorded every two or three days (depending on the weather), during the main flowering season of ragweed (August and September).

For the same period, the daily ragweed pollen concentration over Montréal was estimated with a Burkard™ recording spore trap (Burkard Manufacturing Co., Ltd., Rickmansworth, U.K.) located on top of a 4-story building on the University of Montréal campus. This has been the reference station for aerobiological data at Montréal since 1985. Pollen was captured (over 24-h periods) directly onto a 75×25 mm microscopic slide, coated with glycerin-gelatin adhesive (49 g glycerin, 7 g gelatin, 28 ml water). Slides were replaced every 24 h at 9:00 a.m., local time. The slide passes under the 2×14 mm air intake slit at 2 mm/h. The 48×14 mm trace on each exposed slide was over laid with a mixture of gelatin and basic fuchsin before being covered with a 50×20 mm cover slip. Pollen grains were counted and identified on ten 48 mm horizontal lines, having the width of the microscopic field at 400X, regularly spaced along the exposed surface. Using the area of the exposed surface, the area scanned by the 40X objective, and the flow rate of the sampler (10 l/min), the number of pollen grains per cubic meter of air for each 24-h period was calculated.

Because pollen concentrations were not normally distributed, nonparametric statistics were used for data analysis. The Kolmogorov-Smirnov (K-S) test is a nonparametric test used for comparisons of two independent samples that uses the cumulative distribution of observations. The Kruskall-Wallis (K-W) test is a nonparametric test used for the comparison of independent samples, similar to a one-way analysis of variance and uses ranks of observations. Both the K-S and K-W tests indicate whether independent samples have been drawn from populations with the same distribution.

IV. RESULTS

A. 1990
Figure 3 shows airborne ragweed pollen concentrations (aerobiological data) and mean category of open flowers for all hierarchical levels of flowers

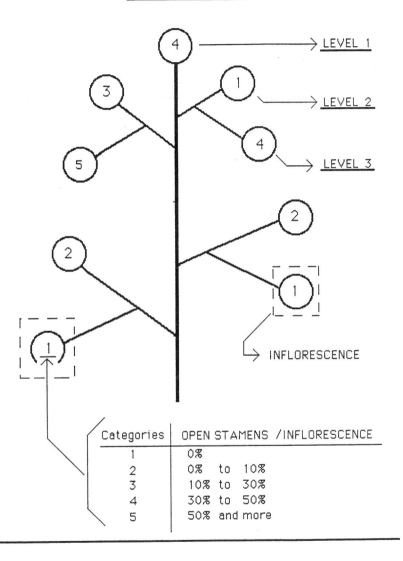

Diagram of Ragweed
flowering architecture

Categories	OPEN STAMENS /INFLORESCENCE
1	0%
2	0% to 10%
3	10% to 30%
4	30% to 50%
5	50% and more

FIGURE 2 Ragweed flowering architecture and the three hierarchical sampling levels used for the study of ragweed phenology.

(phenological data). Aerobiological and phenological data sets were compared within three parts of the ragweed season: phase 1 (15 to 24 August); phase 2 (25 August to 5 September); and phase 3 (6 to 17 September). Statistically, the aerobiological and phenological distributions are different within each of these phases. However, there is a decrease in the probability value (extracted from Z values of a K-S test) from the first and second phases (both with a $p = 0.01$), to the third phase (September 6th to 17th) ($p = 0.005$) meaning that comparing aerobiological and phenological data sets from the three phases, those from the third are most different.

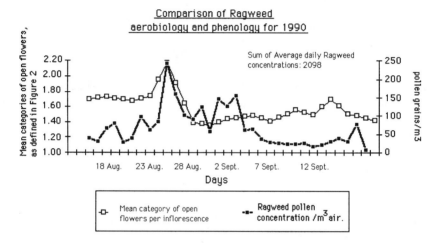

FIGURE 3   Ragweed aerobiology and phenology for 1990.

If we analyze open flower categories by hierarchical levels (Figure 4) and compare these to the pollen curve (shown in Figure 3), we can see that only open inflorescences at levels 1 and 2 (i.e., apical flowers) correspond to the onset of ragweed pollen dispersal. However, flowers at level 3 (i.e., lateral flowers) appear to affect the pollen maximum (i.e., the beginning of dehiscence of stamens from level 3 corresponds to the onset of pollen maximum).

Once past the pollen maximum the mean category of open flowers for all hierarchical levels is relatively stable and appears unrelated to airborne pollen concentrations. A K-W test confirms the similarity of all three levels at this stage with a calculated H of 0.95 (the critical value at $p = 0.05$ being 5.99).

B.   1991

In 1991, pollen concentrations were not obviously related to the category of open flowers (Figure 5). Onset and maximum anthesis, as determined by phenological data, exhibited very well-defined peaks, but little airborne pollen seems to have been released at this time. It was only after peak anther dehis-

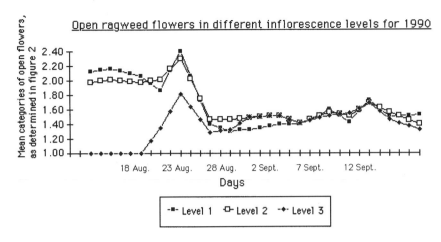

FIGURE 4 Ragweed phenology by hierarchical level of flowers (see Figure 2) for 1990.

cence (maximum mean category of open flowers) was past (after August 24th) that we saw a sharp increase in the airborne pollen concentrations. At this stage, field data revealed a constant situation of few open flowers still releasing pollen. Statistical comparisons consistently indicated significant differences between aerobiological and phenological data sets both before ($p = 0.001$) and after ($p = 0.006$) maximum anthesis.

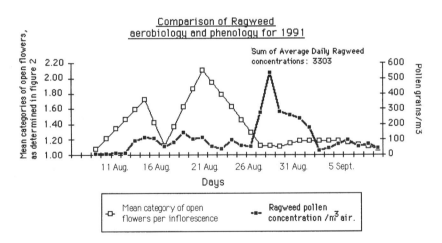

FIGURE 5 Ragweed aerobiology and phenology for 1991.

Graphing the category of open flowers by the different hierarchical levels in 1991 (Figure 6) shows the same general behavior as in the previous year, i.e., levels 1 and 2 have open stamens before level 3, but all levels follow the same trend after the flowering maximum. However, statistically the three

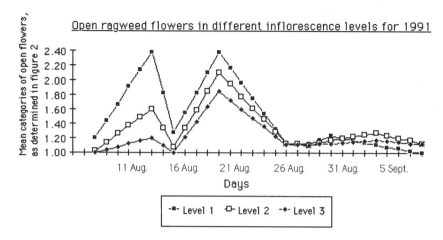

FIGURE 6   Ragweed phenology by hierarchical level of flowers (see Figure 2) for 1991.

hierarchical distributions are different, even after the pollen maximum (K-W test; H = 6.47).

V. DISCUSSION

For the 1990 data set some agreement could be seen between aerobiology (airborne pollen) and phenology (open flowers) of ragweed; this was not the case in 1991. Weather differences might explain the difference between these two ragweed seasons. The ragweed pollen output was 60% higher in 1991 when compared to 1990. Cumulative heating degree-days (>5°C) (Figure 7) could partially explain the high pollen production in 1991.[7] Indeed, the spring of 1991 was unusually warm (see May in Figure 7). Mean temperatures during ragweed pollen season also reflect this warmer trend the second year (15.6°C in 1990; 16.8°C in 1991).

Incidents of precipitation occurred only before or after the main period of ragweed flowering in both years (19.8 mm on 18 August, 9.2 mm on 27 August in 1990; 21.6 mm on 7 August, 32.4 mm on 31 August in 1991) and were unlikely to have affected either flowering or pollen prevalence in air over the season. Relative humidity in eastern Canada is too constant from year to year to be of any predictive value (87% average max, 53% average min for 1990; 88% average max, 54% average min for 1991).

In 1990, there was good agreement between pollen concentrations and flowers opening before August 26 for all three levels of inflorescence (see Figure 3). However, in 1991, maximum airborne pollen is found well after the peak anthesis (Figure 5).

Site-to-site differences may be responsible for the difference in emission and dispersal between the two years. Different sites were used for phenology

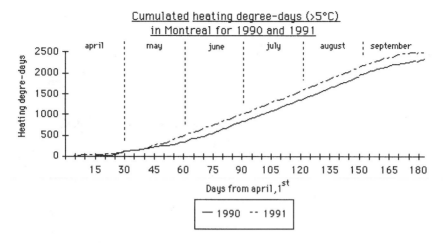

FIGURE 7 Cumulated heating degree-days over 5°C for 1990 and 1991.

in 1990 and 1991. The 1990 site, which showed closer correlation between flowering phenology and airborne pollen levels, was also the nearest to the pollen station. We have already shown that a 20-m high pollen station could represent pollen flux over at least 250 km², but this applied to symptomatology not to phenology.[1] In 1991, maximum wind speeds averaged 33 km/h from August 26 to 31, while they averaged only 20 km/h the previous weeks, and 19 km/h during similar periods in 1990 (in August). It is possible that the large increase in pollen concentration after anthesis was caused by release of residual pollen from dehiscent flowers, refloating of settled grains, and long distance transport, all associated with higher wind speed.

VI. CONCLUSIONS

Patterns of ragweed pollen prevalence (ambient levels) may be related to the hierarchical level of inflorescences releasing pollen on individual plants. Flowers at levels 1 and 2, located on branch tips, release pollen before level 3, located at branch nodes. Opening of flowers at levels 1 and 2 can be linked to the onset of the airborne pollen season, while emission at level 3 contributes to maximum airborne levels.

Average temperature during the growing season appears to affect the hierarchical distribution of flower opening. Release and dispersion may also be related to wind speed.

This small data set presents some clues as to the relationships between patterns of ragweed flowering and airborne concentrations of ragweed pollen. We hope presentation of these preliminary results will stimulate further research in the fascinating field of aerobiology.

REFERENCES

1. Comtois, P. and Gagnon, L., Concentration pollinique et fréquence des symptômes de pollinose: une méthode pour déterminer les seuils cliniques, *Rev. Fr. Allergol.,* 28(4), 279, 1988.

2. Sherknies, D., Un modèle stochastique de prévisions polliniques, M.Sc. Thesis, Département de Géographie, Université de Montréal, 1990, 83.

3. Farnham, J.E., Mason, D., Batchelder, G.L., and Colby, F.P., Ragweed pollen forecasting, in *Aerobiology, Health and Environment,* Comtois, P., Ed., Université de Montréal, 1989, 15.

4. Harrington, J.B., Atmospheric diffusion of ragweed pollen in urban areas, Ph.D. Thesis, University of Michigan, Ann Arbor, 1965, 382.

5. Payne, W.W., Air pollution by ragweed pollen II: the source of ragweed pollen, *J. Air Pollut. Control Assoc.,* 17, 635, 1967.

6. Solomon, W.R., Air pollution by ragweed pollen IV: aspects of human response to ragweed pollen, *J. Air Pollut. Control Assoc.,* 17, 656, 1967.

7. Comtois, P., Batchelder, G.L., and Sherknies, D., Pre-season pollen forecasting, in *Aerobiology, Health and Environment,* Comtois, P., Ed., Université de Montréal, 1989, 1.

Chapter 3

THE INFLUENCE OF WIND SPEED ON THE AMBIENT CONCENTRATIONS OF POLLEN FROM GRAMINEAE, *PLATANUS*, AND *BETULA* IN THE AIR OF LONDON, ENGLAND

Jean C. Emberlin
Jane Norris-Hill

CONTENTS

I. ABSTRACT

Data from a monitoring station in northcentral London are used to investigate the influence of wind speed on ambient concentrations of pollen from three allergenic taxa over the three years, 1987–1989. The analysis considers three contrasting source types, namely Gramineae which originates mostly from outside the conurbation, *Betula* which derives partly from outside the city but has some local sources, and *Platanus* which is almost exclusively local in origin. Standardized pollen counts for dry days are examined in relation to wind speed regimes within categories of temperature ranges, synoptic situations, and sunshine hours. Significant relationships were found between wind speed and concentrations of *Betula* and Gramineae, but not of *Platanus* pollen.

II. INTRODUCTION

Wind exerts an important influence on the dispersal of all anemophilous pollen but little has been done to elucidate its role in medium- or long-range transport. One of the main difficulties in examining the influence of wind speed on pollen concentrations is to separate its effects from those of other meteorological variables such as relative humidity, precipitation, and temperature. There is also the problem of considering wind speed as a variable on its own rather than as one component of a group of factors which govern air movement, including turbulence, mixing depth, and atmospheric stability.

Despite these and other problems, many aerobiologists have attempted to determine the role of wind speed within broad analyses of the effects of meteorological variables, and several have reported significant relationships between wind speed and the daily pollen concentrations. For instance, Ljunkuist et al.,[1] found that wind speed was the most important variable influencing daily grass pollen concentrations during a 4-year study at three stations around Stockholm, and McDonald[2] reported a relationship between pollen concentration peaks and high wind speeds in Galway, Ireland. Mandrioli et al.,[3] described the importance of wind speed and gustiness on the resuspension of pollen grains, and examined the role of those variables on pollen dispersal from its source. In addition, many workers have cited evidence for the importance of wind speed in the dispersal and transport of fungal spores.[4,5]

The studies reported here examine the relationship between wind speed and airborne concentration of pollen from three kinds of sources, seeking to identify the potential contribution of wind speed for predictive modeling.

III. STUDY AREA AND METHODS

The study was conducted in Islington, in northcentral London, an area of flat or gently undulating topography without abrupt changes of slope. The only notable landscape feature is found in the northern part of the borough where Highgate Hill rises to about 80 m above the general level of the terrain. The urban structure consists of a mixture of low- and medium-rise buildings, and occasional tower blocks, interspersed with gardens and larger open spaces such as parks and cemeteries. Street widths vary from four-lane major roads to narrow residential streets.

Pollen concentrations were monitored using a Burkard 7-day volumetric spore sampler (Burkard Manufacturing Co., Ltd., Rickmansworth, U.K.) sited on the flat roof of a 6-story building 25 m above street level within an area of predominantly low-rise buildings. Pollen counts for the three taxa (Gramineae, *Betula,* and *Platanus*) were taken from contiguous 4 mm broad transects on the daily slides, corresponding to 2-h intervals. These counts were averaged to obtain daily concentrations.

Meteorological data for nine parameters were obtained from the London Weather Center, situated 3 km to the south of the pollen trap. The London Weather Center anemometer is mounted considerably above the general roof level at about 10 m above ground level on State House, in High Holborn. In a 6-month study comparing the anemometer for this site and a nearby site, Lee[6] concluded that the wind record from the London Weather Center represented the gross airflow over the city rather than local conditions influenced by street orientation or building density.

The pollens chosen for study were Gramineae (grass, about 20–60 μm in diameter), which originates mostly from outside London, the main source areas being about 20 miles away; *Betula* (birch, about 25–30 μm in diameter) which is derived mostly from the surrounding rural areas but does have some local sources in parks and gardens; and *Platanus* (plane, about 18–22 μm in diameter) which is predominantly local in origin. It was suspected that the role of wind speed in influencing the dispersal of pollen from these three contrasting sources could be very different.

The annual pollen seasons for each of the three taxa have been defined as the periods during which 90% of the pollen catch occurred. The start of the season is the day when the cumulative sum reaches 5% of the seasonal total, and the season ends on the day when the 95% level has been reached.[7] The problem of variability in pollen production between and within seasons has been overcome by standardizing the counts following the method of Moseholm et al.,[8] whereby daily counts are multiplied by 1000 and divided by the cumulative yearly pollen count. The daily counts have also been $\log_{10}$ transformed to achieve normalization.

The records for the three taxa were subjected to the same range of analyses considering minimum, average, and maximum wind speeds but this chapter will concentrate on the most significant results. Analyses with $\log_{10}$-transformed wind speed data were conducted but did not improve the significance levels obtained.

IV. RESULTS AND DISCUSSION

Hirst,[9] estimates that the efficiency of the Burkard trap averages 80% in normal wind conditions for particles of pollen size (10–50 μm) but it is acknowledged that this varies in high wind velocities,[9,10] as greater volumes of air may be forced through the orifice.

A. GRAMINEAE

During the Gramineae pollen seasons of 1987–89 the frequencies of daily average wind speeds were fairly evenly spread, with a slight peak in the 3.1–4 m/sec range (Figure 1). Daily minimum wind speeds were mostly in the 0–3 m/sec range (120 d); daily maximum speeds peaked in the 6.1 to high range (84 d).

In a comparison of the magnitude of correlation coefficients between the average pollen counts and the meteorological variables, average wind speed ranked fourth in significance with maximum temperature, rainfall, and average humidity (Table 1) being more strongly related. The relationship between wind speed and average pollen counts shows a significant negative correlation. However, partial correlation coefficients are significant for minimum wind speed on dry days (Table 2), indicating a positive relationship when controlling for average humidity, maximum temperature, average cloud cover, and sunshine hours. This suggests that a minimum critical wind speed may exist to affect transport of the grass pollen into the city. During some anticyclones, convection currents will generate local circulation within the urban area.[11,12] This city thermal wind system, which exists with large scale stagnation, is largely self-contained, having a diurnal reversal of flow.[13] Pollen counts tend to be higher when a regional airflow of low to moderate speed operates. The greatest potential for transport often exists with weak winds as both horizontal dispersal and turbulent diffusion are curtailed.

Significant differences occurred between the wind speed regimes recorded on days of low pollen counts (<50 grains/m³) and those on days of high pollen counts (Table 3). Adopting this approach, low pollen counts were associated with higher wind speeds. Greater wind speeds in the urban area would enhance the role of turbulence and advection. Wind speed governs the amount of forced convection generated in the boundary layer due to internal shearing between air layers. It is evident that the greater turbulent activity at higher wind speeds leads to rapid dilution of pollen concentrations. This is especially pertinent in

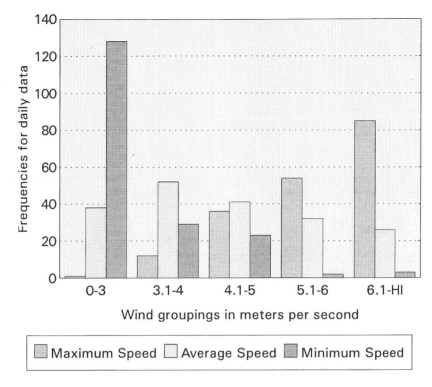

FIGURE 1 Frequency of maximum, average, and minimum daily wind speeds, grouped by five wind speed ranges during the Gramineae pollen seasons 1987–89.

TABLE 1   Correlation Coefficients for the Relationship between Average Daily Gramineae Pollen Concentrations and the Main Meteorological Variables

Model: Logged and Standardized Counts

Pearson Product Moment Correlations

	Rainfall	Maximum Temperature	Average Humidity	Average Cloud Cover	Sun	Average Wind
r	0.45	0.55	–0.56	–0.40	0.40	–0.21
p	0.011	0.000	0.006	0.306	0.461	0.019

Note: r = correlation coefficient; p = probability level.

the transition of the airflow from the rural to the urban environment because the depth of the boundary layer will be increased by turbulence induced by the surface roughness of the city. Although wind speeds will be decreased by the greater frictional drag resulting in some convergence, the overall effect will be to increase mixing and dilution.

**TABLE 2   Partial Correlation Coefficients for the Relationship
between Average Daily Gramineae Pollen Concentrations
and the Main Meteorological Variables Considering
Minimum and Maximum Wind Speed Regimes**

Daily Grass Pollen 1987–1989

Wind Speed Aspect				
Minimum		Maximum		
r	p	r	p	Controlling for:
0.08	0.181	−0.08	0.024	RH
0.21	0.025	0.17	0.44	RH, max. temp.
0.21	0.025	0.07	0.44	RH, max. temp., avg. cloud, sunshine hours

Note: Dry days only: r = partial correlation coefficient; p = probability level; RH = relative humidity.

**TABLE 3   Daily Average Gramineae Pollen Counts
1987–89. Analysis of Variance between
Wind Speed Regimes on Days with Low
Counts and Those with High Counts**

Group	Mean	F value	Probability
Maximum Wind Speed			
1	5.81	1.90	0.004
2	6.73		
Average Wind Speed			
1	3.97	1.85	0.005
2	4.79		
Minimum Wind Speed			
1	2.19	1.58	0.036
2	2.91		

Note: Group 1, days with 50 or more grains/m³; Group 2, days
with less than 50 grains/m³ air. Wind speeds are in meters
per second (m/s).

In a further examination of this aspect, daily grass pollen counts were analyzed in relation to the synoptic situation. High and low pressure systems occurred in approximately equal frequencies (Figure 2). Depressions were divided according to the location of the low pressure center relative to the site because their associated airflows possess different characteristics, especially of turbulence. For example, airflows associated with systems centered to the west of London tend to be less stable than those centered to the east. Significant differences were found between daily grass pollen counts grouped by the location of the low pressure systems (p <.005) but these may result from differences in the amount and type of rainfall occurring as this was not included in the analysis.

FREQUENCY OF SYNOPTIC SITUATIONS
GRASS POLLEN SEASONS 1987-89

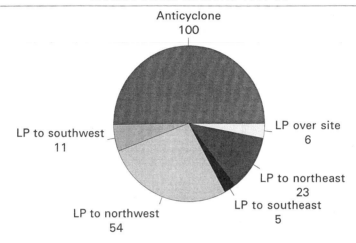

LP low pressure

FIGURE 2 Frequency of synoptic situations during the grass pollen seasons 1987–1989 (LP = low pressure).

B. *BETULA*

Typically, the seasons for *Betula* pollen (mostly from *Betula pendula* and *Betula pubescens*) in London range from late March to late May. During the 1987–89 seasons the most frequently occurring average wind speed regime was between 3.1–4.0 m/sec (Figure 3), with minimum winds on most days (52) being below 3 m/sec and maximum winds on 35 days exceeding 6.1 m/sec..

In the analysis of correlations between daily *Betula* pollen counts and the meteorological variables (average, minimum, or maximum), wind speed was not significant. Similarly, it did not feature as significant in the partial correlations, controlling for multiple variables. Analysis of variance between the daily average pollen concentrations recorded on days grouped by maximum wind speeds yielded a significant difference ($p = 0.05$) indicating that pollen concentrations were limited by high wind velocities. Similarly, a significant relationship with minimum wind speed occurred in the daily pollen data grouped into high (>50 grains/m^3) and low pollen days (Table 4). Wind speeds on days with high pollen concentrations were significantly less than those on

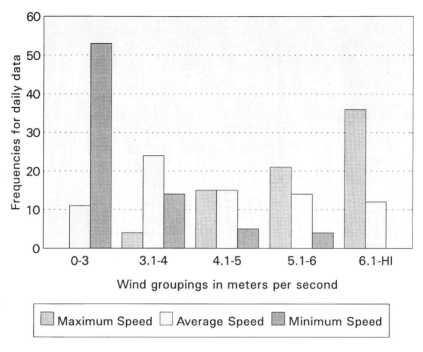

FIGURE 3 Frequency of maximum, average, and minimum daily wind speeds, grouped by five wind speed ranges during the *Betula* pollen seasons 1987–89.

TABLE 4 Results of the Analysis of Variance between Wind Speed Regimes Recorded on Days with High Concentrations of *Betula* pollen and Those on Days with Low Concentrations

Group	Mean	F value	Probability
Maximum Wind Speed			
1	6.5	1.23	0.582
2	6.5		No significant difference
Average Wind Speed			
1	4.5	1.73	0.582
2	4.7		No significant difference
Minimum Wind Speed			
1	2.14	2.7	0.008
2	3.00		Significant difference

Note: Group 1, days with 50 or more grains/m^3; Group 2, days with less than 50 grains/m^3 air. Wind speeds are in meters per second (m/s).

low pollen days. This follows the results obtained in the analysis of the grass data and similar mechanisms are likely to be involved.

Anticyclones were less frequent in the *Betula* seasons than in those for grass. The most frequent low pressure systems were those centered to the southwest. These are likely to produce unsettled, showery weather. No significant differences were detected for *Betula* pollen concentrations in relation to synoptic type.

Overall, wind speed appears to be less important as a controlling variable on daily average *Betula* pollen concentrations than in the case of Gramineae, but some significant relationships have been identified. The interpretation of the dispersal of *Betula* pollen is complicated by the dual nature of the source areas. Previous work on the spatial variation of ambient *Betula* pollen in the area indicated that a substantial proportion of this pollen was imported from rural areas.[14] Steep concentration gradients occurred around local sources, suggesting that most of the pollen released within the urban canopy layer was removed fairly rapidly by dry deposition and filtration. The relative contributions of near and distant sources cannot be identified in the current data but the influences of wind speed on the ambient concentrations at roof level would likely be different in each case.

C. PLATANUS

The seasons for *Platanus* pollen in London typically extend from early April to late May or early June. During these times for the period 1987–89, the wind speed profile (Figure 4) was markedly different from those for Gramineae and *Betula*. This is particularly noticeable in the daily maximum wind speeds which occurred more frequently in the 3.1–4.0 m/sec group than in other wind speed groupings.

Although relatively small (22 μm diameter), *Platanus* pollen is heavy, having a specific gravity of 0.92.[15] The main source is the plane trees which line many streets within the urban canopy layer while, during the current study, the pollen was collected at roof height (that is, above the canopy). Wind speeds at street level are often lower than those in the urban boundary layer, depending of course, on factors such as street orientation, street dimensions, and atmospheric stability. No significant relationships were found between daily *Platanus* pollen concentrations and wind speed. In the examination of synoptic situations, the frequency of systems was similar to that for *Betula* but again, no significant relations emerged.

Evidence from previous work on the vertical distribution of pollen in Islington indicated that *Platanus* pollen was not likely to be dispersed upwards from the urban canopy layer except in very gusty conditions.[16] In these situations, impaction rates would be high, leading to rapid depletion of the pollen from the airflow.

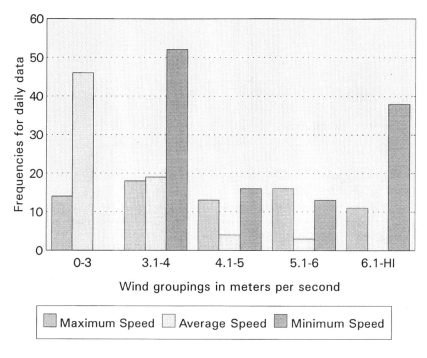

FIGURE 4 Frequency of maximum, average, and minimum daily wind speeds, grouped by five wind speed ranges, during the *Platanus* pollen seasons 1987–89.

V. CONCLUSIONS

Despite the limitations and constraints outlined in the introduction, significant relationships have been identified between wind speed and daily pollen concentrations in the cases of Gramineae and *Betula*. Of these, the role of wind is most clearly apparent in Gramineae and can be incorporated into a predictive model.

A step-wise multiple regression model built on the relationship between daily average Gramineae pollen concentrations, average humidity, maximum temperatures, rainfall, and wind velocity achieves a coefficient of $r = 0.658$. The relationship can be used as the basis for a model to predict daily variations in Gramineae pollen concentrations providing that it is applied in conjunction with an estimate of the potential pollen production for the time within the season.

For *Betula* pollen, significant relationships have been identified with minimum and maximum wind speed. The dual nature of the sources and consequent dispersal of this pollen seem to obscure general relationships. No significant results were obtained in the analyses for *Platanus* pollen concentrations. This is likely to be caused by the limited dispersal of *Platanus* pollen within the urban canopy layer and the contrasts of the wind regime at this level and that at roof height. It is also worth noting that these analyses have not produced any evidence of a significant resuspension of pollen during gusty conditions as has been reported by other researchers.[17]

The experimental design of the study would be improved greatly by including a detailed analysis of regional airflow, together with information about the vertical profiles of wind speed and turbulence. It is hoped that this will be achieved in future work at the site.

ACKNOWLEDGMENTS

Jean Emberlin would like to thank Marion Merrell Dow Pharmaceuticals (UK) for their sponsorship during the preparation of this chapter. Jane Norris-Hill extends thanks to Janssen pharmaceuticals (UK) for supporting her research work.

REFERENCES

1. Ljunkuist S., Bringfelt B., and Fredriksson U., Correlation between the pollen content of the Stockholm air and meteorological data, *Grana,* 16, 145, 1977.
2. McDonald, M., Correlation of airborne grass pollen levels with meteorological data, *Grana,* 19, 53, 1980.
3. Mandrioli, P., Negrini, M. G., and Zanotti, A. L., Mesoscale transport of *Corylus* pollen grains in winter atmosphere, *Grana,* 18, 227, 1980.
4. Gregory, P. H., Distribution of airborne pollen and spores and their long distance transport, 166, 309, 1978.
5. Lyon, F. T., Kramer, C. L., and Eversmeyer, M. G., Variation of air spora in the atmosphere due to weather conditions, *Grana,* 233, 177, 1984.
6. Lee, D. O., Urban influence on wind directions over London, Department of Geography, Birkbeck College, London, 1975.
7. Nilsson, S. and Persson, S., Tree pollen spectra in the Stockholm region Sweden, 1973–1980, *Grana,* 20, 179, 1981.
8. Moseholm, L., Weeke, E., and Petersen, B. N., Forecast of pollen concentrations of Poaceae (grasses) in the air by time series analysis, *Pollen et Spores,* 19, 305, 1987.
9. Hirst, J. M., The capabilities and limitations of the Hirst spore trap, *Acta Allergol.* (Supp. 7), 149, 1960.
10. Ogden, E., Ogden, C., Raynor, G. S., Hayes, J. V., Lewis, D. M., and Haines, J. H., *Manual for Sampling Pollen*, MacMillen, C., 1974.
11. Chandler, T., *The Climate of London*, Hutchinson, London, 1965.
12. Bornstein, R. D., Reported observations of urban effects on winds and temperature, in and around New York City, reprints of Am. Meteor. Soc. Conf. on Urban Environ., Philadelphia, PA, Oct. 31–Nov. 2, 1972.
13. Oke, T., *Boundary Layer Climates*, Methuen & Co., 1978.
14. Emberlin J. and Norris-Hill J., Spatial variation of pollen abundance in North-Central London, *Grana,* 30, 190, 1991.

15. Waldbot, B., *The Health Effects of Environmental Pollutants*, C. V. Mosby Co., St. Louis, MO, 1973.
16. Bryant, R. H., Emberlin, J. C., and Norris-Hill J., Vertical variations of pollen abundance in North-central London, *Aerobiology*, 5, 123, 1989.
17. Tampieri, F., Medium range transport of airborne pollen, *Agric. Meteorol.*, 18, 19, 1977.

Chapter **4**

AEROPOLLEN OF MIMOSOIDEAE

Inés Hurtado
Julio Alson

CONTENTS

39

I. ABSTRACT

Aerobiological surveys carried out in a neotropical location (Caracas, Venezuela) have a relatively large representation of compound pollen grains which are typical of the Mimosoideae. In this group, as in others of the legume family, anemophily has been rarely reported. We have studied the compound grains in slides and tapes exposed at the sampling location, as well as their sources in field and herbarium collections. As a result of these studies seven airborne pollen types of the Mimosoideae genera *Acacia, Calliandra, Mimosa,* and *Piptadenia* are described and illustrated.

II. INTRODUCTION

Compound pollen grains typical of the Mimosoideae are rarely recognized in aerobiological surveys from the temperate zones. In contrast, they seem relatively frequent in our neotropical location (Caracas, Venezuela),[1,2] a circumstance which has stimulated our investigation of these compound airborne units and their sources. Although pollination in the Mimosoideae, a subfamily of the Leguminosae, is commonly zoophilous, anemophily has also been described in this large group of plants.[3] This paper describes seven pollen species from the Mimosoideae genera *Acacia, Calliandra, Mimosa,* and *Piptadenia*. Records derived from volumetric air sampling illustrate the annual dispersal patterns of the more frequently recovered *Mimosa* grains.

III. MATERIALS AND METHODS

A. AIR SAMPLING

For the past several years continuous air sampling has been conducted in Caracas, Venezuela (1500 m alt., 10° 30′ N lat., 66° 70′ W long.) using a Durham collector,[4] and a Rotorod sampler (Aeroallergen Model, Sampling Technologies, Minnetonka, MN). The Rotorod was calibrated to sample for 60 sec every 10 min and rotate at 2400 rpm. Both instruments were operated at 10 m above ground, on the open rooftop of a 4-story building. Slides and rods were covered with a thin film of silicone grease (Dow Corning, USA), and after exposure stained with basic fuchsin (Calberla's solution[5]) and read at a magnification of 400×. Pollen counts/m^3 of air were derived from readings of the entire surface of the two 24-h exposed rods.

B. POLLEN AND SOURCE IDENTIFICATION

Pollen identification was aided by reference to the literature,[6–9] and to reference slides prepared from herbarium and field specimens. Herbarium specimens were from the Universidad Central de Venezuela at Maracay Herbarium. Fresh specimens were collected within the several hectares of vegetation which surround the sampling site or in neighboring localities a few kilometers away. Anther material was shaken onto a drop of fuchsin-stained glycerin jelly on a heated slide, covered, and sealed.[5] Voucher specimens and reference slides have been filed in the collection of the laboratory.

Photographs were taken with the Nikon Optiphot Microscope and Microflex HFX photomicrographic attachment (Nippon Kogaku K.K, Japan).

IV. RESULTS

A. POLLEN DESCRIPTION AND SOURCES

The compound grains recovered on air samplers are illustrated and described (Figures 1–7, Table 1). They are 4- to 16-celled units which vary greatly in size, from 9 to 156 μm; in shape, from nearly spherical to highly irregular; and in the way the single grains are arranged, either tightly packed as in *Mimosa albida/pudica* (Figure 3) or loosely aggregated as in *Piptadenia pittieri* (Figure 6).

Calliandra spp. are cultivated as ornamentals, all other Mimosoideae are weedy plants. *Acacia* spp., *Calliandra* spp., *Piptadenia* spp., and *M. arenosa* are trees. Other *Mimosa* spp. are shrubs with *M. pudica* being the predominant species of the area.

B. POLLEN COUNTS

Mimosoideae pollen grains were recovered year-round on air samples. *Calliandra* spp. were recovered only on Durham samples.

Monthly sums of average daily *Mimosa* pollen counts/m^3 of air in 1987–1988 are shown (Figure 8). Highest levels of *M. albida/pudica* pollen recovery occurred from September to December and of *M. arenosa* pollen recovery during October. During peak months, daily concentrations as high as 10 grains/m^3 air of *M. albida/pudica* pollen, and as high as 7 grains/m^3 air of *M. arenosa* pollen, have been recorded.

V. DISCUSSION

Compound grains are characteristic of, and predominant in the Mimosoideae,[8] and are easily differentiated from single grains and individually from each other as the microphotographs illustrate (Figures 1–7). In our area, the only non-Mimosoideae compound grain we occasionally recover from air is a large Ericaceae tetrad illustrated in a previous publication[10] which is differentiable from the small and medium size tetrads described here.

The Mimosoideae grains we describe are highly diverse. Both the smallest (9 μm *Mimosa*) and the largest (156 μm *Calliandra*) grains of our pollen

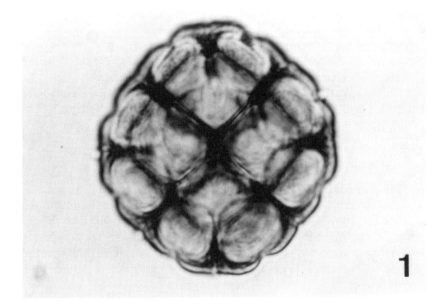

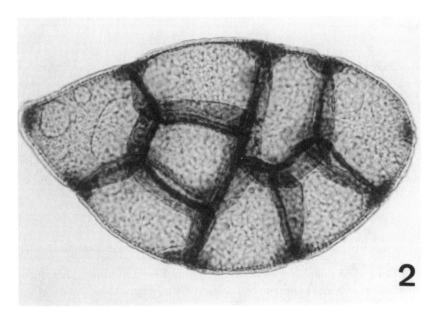

FIGURES 1–2 1, *Acacia macracantha* (original magnification, 400×). **2,** *Calliandra falcata* (original magnification, 200×).

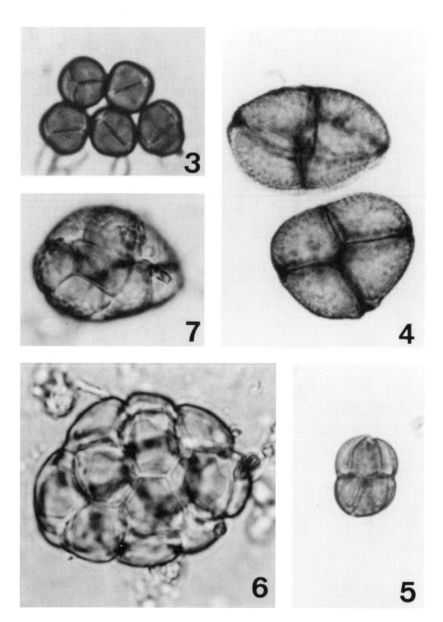

FIGURES 3–7 **3,** *Mimosa pudica.* **4,** *M. spiciflora.* **5,** *M. arenosa.* **6,** *Piptadenia pittieri.* **7,** *P. robusta.* (Original magnification 1000×).

spectrum belong in the group. Note that the small *Mimosa* may escape detection by the untrained eye which often scans for much larger particles. Its concentration is probably underestimated by the Rotorod sampler which is most efficient for particles above 20 μm.[11] Among 82 species from India, *M. pudica*, which had the smallest anther and grain, ranked fourth in pollen production.[12]

TABLE 1   Airborne Mimosoidae Pollen

Figure Number	Taxon	Number of Cells	Shape	Surface[a]	Apertures[b]	Size (μm)[c]
3	*M. albida/pudica*	4	Spheric, regular tetahedral tetrad	S	+	9 ± 1
4	*M. spiciflora*	4	Heteropolar, irregular tetragonal tetrad	S	+ Thickened	26 ± 1
5	*M. arenosa*	8	Oval, regular polyad (two inverted tetrads)	O	+	14 ± 1
2	*Calliandra falcata/schultzei*	8	Heteropolar, irregular polyad (highly disymmetric, tetrads undistinguishable)	O	+	156 ± 4
7	*Piptadenia robusta*	12	Heteropolar, irregular polyad (3 tetrads, 2 similarly oriented)	O	+	27 ± 2
6	*P. pittieri*	16	Roughly oval, irregular polyad (tetrads undistinguishable, loosely aggregated)	S	+ Aspidotes	43 ± 2
1	*Acacia macracantha*	16	Spheric, regular polyad (4 radially arranged tetrads)	S	+	42 ± 3

[a] S = smooth, O = ornate
[b] + = apertures:simple, porate
[c] Average of the largest pollen diameter in 10 compound grains measured ± s.e.m.

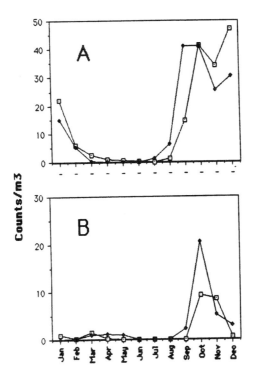

FIGURE 8 Seasonal pattern of airborne *Mimosa* pollen, monthly sums of daily pollen grains/m³ of air recovered by Rotorod sampler. A = *Mimosa albida/pudica*; B = *Mimosa arenosa;* □ = 1987; ♦ = 1988.

In our pollen spectrum of approximately 60 grain types[2] the Mimosoideae are well represented with seven types. They contribute an estimated 5% of the total annual pollen counts and are the fourth largest contributor of indigenous airborne pollen behind the more prolific pollen shedding species of *Cecropia*, Urticales, and *Acalypha*.[2]

Anemophily of *Acacia* spp. has been reported[13–20] and the *Acacia* pollen is a popular example of airborne compound grains in aerobiology manuals.[5,21] Anemophily of *Mimosa* spp., which was early recognized in our Laboratory,[16] has been reported in two other tropical countries, the Phillipines[22] and Thailand.[20] In these countries, the taxon is the first and second largest contributor, respectively, of airborne pollen. *Calliandra* grains, but not *Piptadenia* grains, have been previously reported as recovered from air.[16]

In contrast to the common compound grains, single grains of Mimosoideae are virtually absent from our air samples. In the single grain category, *Prosopis* spp. are the most often cited as anemophilous.[5,15,22–24] Only one species of this genus has been described in Venezuela[6] and we have yet to recover its pollen in local surveys.

The allergenicity of *Acacia*,[25-27] *Mimosa*,[22] and *Prosopis*[24,28-29] has been documented. Sensitization to *Acacia* represents an occupational hazard for floriculturists with increases in sensitization rates with increased exposure.[30]

Because of the consistent presence and peak concentrations of Mimosoideae pollen in our air, their inclusion in epidemiological and clinical surveys of allergic respiratory conditions is recommended.

ACKNOWLEDGMENTS

Henry Ramos provided valuable assistance with the counts of pollen grains. Herbarium specimens were generously supplied by Dr. Lourdes Cárdenas de Guevara (Universidad Central de Venezuela, Maracay, Venezuela).

REFERENCES

1. Hurtado, I. and Riegler-Goihman M., Air sampling studies in a tropical area (Venezuela). Frequency and periodicity of pollen and spores, *Allergol. et Immunopathol.*, 12, 449, 1984.
2. Hurtado, I. and Alson, J., Air pollen dispersal in a tropical area, *Aerobiologia*, 6, 122, 1990.
3. Kalin Arroyo, M. T., Breeding systems and pollination biology in Leguminosae, in *Advances in Legume Systematics*, Polhill, R.M. and Raven, P.H., Eds., 1981, 723.
4. Brown, G. T., *Pollen-Slide Studies*, Charles C Thomas, Springfield, IL, 1949.
5. Ogden, E. C., Raynor, G. S., Hayes, J. V. and Haines, J. H., *Manual for Sampling Airborne Pollen*, Hafner Press, New York, 1974.
6. Cárdenas-Guevara, L., Los géneros venezolanos de las Mimosoideae (Leguminosae), *Rev. Fac. Agron. (Maracay)*, 7, 109, 1974.
7. Caccavari, M. A., Estudio de los caracteres del polen en las Mimosa-Lepidotae, *Pollen et Spores*, 28, 29, 1986.
8. Guinet, Ph., Mimosoideae: the characters of their pollen grains, in *Advances in Legume Systematics*, Polhill, R.M. and Raven, P.H., Eds., 1981, 835.
9. Salgado-Labouriau, M. L., *Contribuçao a Palinologia dos Cerrados,* Editora Academia Brasileira de Ciencias, Rio de Janeiro, Brazil, 1973.
10. Hurtado, I., Leal-Quevedo, F. J., Rodríguez-Ciodaro, A., García-Gómez, E., and Alson-Haran, J. A., A one year survey of airborne pollen and spores in the neotropical city of Bogotá (Colombia), *Allergol. et Immunopathol.*, 17, 95, 1989.
11. Solomon, W. R. and Mathews, K. P., Aerobiology and inhalant allergens, in *Allergy: Principles and Practice*, Middleton, E., Reed, Ch.E., and Ellis, E.F., Eds., C.V. Mosby, St. Louis, MO, 1983, 1143.
12. Subba Reddi, C. and Reddi, N. S., Pollen production in some anemophilous angiosperms, *Grana*, 25, 55, 1986.
13. Sánchez-Medina, M. and Fernández, A., Allergenic pollens in Bogotá, Colombia, South America, *J. Allergy*, 38, 46, 1966.
14. Lewis, W. H. and Vinay, P., North American pollinosis due to insect-pollinated plants, *Ann. Allergy*, 42, 309, 1979.
15. Lewis, W. H., Airborne pollen of the Neotropics, *Grana*, 25, 75, 1986.
16. Hurtado, I. and Riegler-Goihman, M., Air sampling studies in a tropical area. I. Airborne pollen and fern spores, *Grana*, 25, 63, 1986.
17. Singh, B. P., Singh, A. B., Nair, P. K., and Gangal, S. V., Survey of airborne pollen and fungal spores at Dehra Dun, India, *Ann. Allergy*, 59, 229, 1987.

18. Al-Eisawi, D. and Dajani, B., Airborne pollen of Jordan, *Grana*, 27, 219, 1988.
19. Romano, B., Pollen monitoring in Perugia and information about aerobiological data, *Aerobiologia*, 4, 20, 1988.
20. Dhorranintra, B., Limsuvan, S., Kanchanarak, S., and Kangsakawin, S., Aeroallergens in northern and southern provinces in Thailand, *Grana,* 30, 493, 1991.
21. Smith, E. G., *Sampling and Identifying Allergenic Pollens and Molds*, Blewstone Press, San Antonio, TX, 1984.
22. Cua-Lim, F., Payawal, P. C., and Laserna, G., Studies on the atmospheric pollens of the Philippines, *Ann. Allergy*, 40, 117, 1978.
23. Singh, A. B. and Babu, C. R., Survey of atmospheric pollen allergens in Delhi: seasonal periodicity, *Ann. Allergy*, 48, 115, 1982.
24. Bieberdorf, F. W. and Swinny, B., Mesquite and related plants in allergy, *Ann. Allergy*, 10, 720, 1952.
25. Geller, M. and Rosario, N. A., Skin test sensitivity to *Acacia* pollen in Brazil, *Ann. Allergy,* 180, 1981.
26. Howlett, B. J., Hill, D. J., and Knox, R. B., Cross reactivity between *Acacia* (wattle) and rye grass pollen allergens, *Clin. Allergy*, 12, 259, 1982.
27. Bousquet, J., Cour, P., Guerin, B., and Michel, F. B., Allergy in the Mediterranean area. I. Pollen counts and pollinosis in Montpellier, *Clin. Allergy,* 14, 249, 1984.
28. Shivpuri, D. N. and Parkash, D., A study in allergy to *Prosopis juliflora* tree, *Ann. Allergy*, 25, 643, 1967.
29. Thakur, I. S., Fractionation and analysis of allergenicity of allergens from *Prosopis juliflora* pollen, *Int. Arch. Allergy Appl. Immunol.*, 90, 124, 1989.
30. Ariano, R., Panzani, R. C., and Amedeo, J., Pollen allergy to mimosa (*Acacia floribunda*) in a Mediterranean area: an occupational disease, *Ann. Allergy*, 66, 253, 1991.

Chapter 5

PATHOGENIC AND ANTIGENIC FUNGI IN SCHOOL DUST OF THE SOUTH OF SPAIN

Julia Angulo-Romero
Félix Infante-García-Pantaleón
Eugenio Domínguez-Vilches
Ana Mediavilla-Molina
José M. Caridad-Ocerín

CONTENTS

1-56670-206-2/96/$0.00+$.50
© 1996 by CRC Press, Inc.

I. ABSTRACT

This paper presents a 2-year study of fungi in dust from 12 schools in Córdoba, Spain. Four hundred fifty-six dust samples were collected using a vacuum cleaner and were analyzed culturally for microfungi. Of the 26,338 fungal colonies that were isolated, 38% belonged to potentially pathogenic (capable of causing infectious or hypersensitivity diseases) fungal taxa. The most frequently encountered of the 91 identified taxa were *Alternaria alternata, Aspergillus fumigatus, Aspergillus niger,* and several species of *Penicillium.* In broad terms, most of these exhibited seasonal variations in concentration and were more abundant between April and October. Also, sampling sites often contained characteristic combinations of fungal taxa in their dust.

II. INTRODUCTION

The clinical significance of fungi lies in their potential to induce systemic or localized infections (pathogenic fungi) and hypersensitivity reactions or allergies (antigenic fungi).[1,2] It is becoming evident that all fungi probably have the potential to act as antigens. Most of the infectious fungi and the more thoroughly studied antigenic fungi belong to genera of the Deuteromycotina and Zygomycetes.[3–6]

Fungi give rise to two types of systemic infections depending on the taxa involved; the virulent pathogenic fungi cause diseases such as histoplasmosis, blastomycosis, and coccidioidomycosis, while the opportunistic pathogens especially affect immunologically suppressed individuals, where they induce aspergillosis, zygomycosis, candidiasis, and several other diseases.[1,4] Fungi also induce hypersensitivity reactions or allergies which are probably of greatest social concern.[5,7–9] In principle, all types of spores have the potential to induce a hypersensitivity response in any individual, although the IgE mediated allergies probably occur only in those with the appropriate genetic predisposition. Fungal allergies can be acquired both indoors and out. Fungal allergies acquired indoors may be a serious public health problem[2] and the importance of indoor allergen studies to the accurate diagnosis and treatment of allergies has been pointed out by some authors.[8,10]

As a rule, buildings accommodate a variety of habitats and nutrient sources where fungi can readily grow including wood, food, paper, fabrics, paint, animal and human effluents, and plants. Many of these nutrient sources also become components of dust. Although dust mites clearly contribute a major allergenic component to dust, the fungi and other materials found in dust such as feathers, dermal remains, fabrics, pollens, and insects are also potential or proven allergens.[11–15] Characterizing as wide a spectrum of such allergens as possible could be of great significance to the diagnosis of individual allergies to "house" dust (whether work, school, or domestically derived) and hence to an appropriate choice of the desensitizing allergen to be administered.[16]

In spite of the importance of domestic dust as a source of a host of allergens, it has scarcely been studied with respect to other health effects, including its potential as a source for infectious agents.[17–22] In addition, symptoms such as dry mucosa and skin, erythema, mental fatigue, headaches, recurrent cough and respiratory infections, hoarseness, wheeze, itch and non-specific hypersensitivity, nausea, vertigo, etc. are increasingly reported among nonindustrial workers.[17,18] This group of symptoms, known as the S.B.S. (Sick Building Syndrome),[23] is often encountered in schools and offices character-ized by heavy human traffic and small cleaning budgets and can be worsened by the presence of damp carpets or water damage.[17,18] The assumption is often made that biogenic components of dust contribute to these symptoms.

Inasmuch as humans spend most of their time indoors (up to 80–90% in the case of children, elderly, and sick individuals),[24] nonindustrial habitats (homes, offices, schools, and other shopping and public facilities) should be regarded as possible sites of exposure to dust components. In fact, human traffic results in the transfer, settling, and buildup of dust particles in addition to increased release of the dust and its constituent allergens from the surfaces on which it settles. These factors led us to analyze the culturable fungi in dust from Córdoba schools and classify the dust according to predominant fungal taxa. We also aimed to determine the factors which contribute to contamination in order to facilitate development of efficient prophylactic measures.

III. METHODS

The study was carried out in Córdoba, Spain, a city of 262,000 inhabitants lying about 120 m above sea level, with a Mediterranean climate. Average annual rainfall is 674 mm and average annual temperature is 18°C (average maximum and minimum temperatures are 30°C and 7°C, respectively).

Dust samples were collected at twelve primary education schools, of which six were public and six private. The schools were classified into pub-lic–private pairs according to their location. Each pair was characterized by geographic location in the city, architectonic and urban features, age and occupancy level of the schools, and social status of the pupils. Samples were collected biweekly for two consecutive school years (except during the summer vacation months, July and August) using a customized portable vacuum cleaner (Jumbo Vacuum Cleaner,® U.K. Reg. Des. N° 951869, Hong Kong). At each school the same classroom was always sampled and the floor was vacuumed until an appreciable amount of dust had been collected. Dust samples were stored at room temperature in sterile plastic bags and sealed with adhesive tape for subsequent processing.

Ten milligrams of each dust sample were inoculated onto duplicate Petri dishes containing 2% malt extract agar according to Blakeslee's formula[25] and 5 ml/l of 25% lactic acid in order to acidify the medium and prevent bacterial growth. The dishes were incubated upside-down at 27°C for 7 d, after which colonies were counted, isolated, and identified.

We use the term "pathogenic" in a broad sense to indicate those fungi commonly cited in the literature with a proven capacity to produce some kind of illness (infections and/or antigenic reaction).

IV. RESULTS

A total 26,338 fungal colonies were isolated from the 912 Petri dishes used, giving an average 29 colonies per dish. Colonies numbering 11,980 were isolated in 1985/86, of which 5200 belong to taxa commonly and specifically cited as causing some type of illness; 4848 out of the 14,358 colonies isolated in 1986/87 belong to this same "illness inducing" group. Figure 1 shows the mean number (per 10 mg dust) of fungi commonly considered to cause disease that were encountered throughout the sampling period. As can be seen, the highest mean levels of these organisms were obtained in May and September. Wilcoxon's test was used to compare the number of colonies isolated every month in all of the schools sampled; significant differences were found for September with respect to all other months except May and June. Probability limits are shown in Table 1.

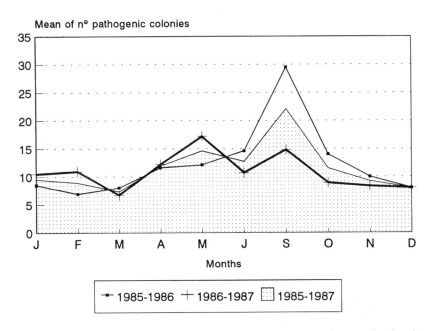

FIGURE 1   Graph showing the average number of pathogenic taxa colonies isolated each month throughout the sampled periods and the average of the two years.

TABLE 1 Wilcoxon Test with the Probability Limits for the Monthly Isolations for All of the Schools

	Sep	Oct	Nov	Dec	Jan	Feb	Mar	Apr	May	Jun
Sep	1.0000									
Oct	0.0164*	1.0000								
Nov	0.0069*	0.6476	1.0000							
Dec	0.0008*	0.3758	0.2192	1.0000						
Jan	0.0018*	0.1209	0.0425*	0.4842	1.0000					
Feb	0.0066*	0.2192	0.1808	0.4839	0.4237	1.0000				
Mar	0.0015*	0.0696	0.0357*	0.1858	0.4938	0.6476	1.0000			
Apr	0.0214*	0.4663	0.5485	0.7751	0.8415	0.4155	0.5296	1.0000		
May	0.3606	0.6682	0.6373	0.2904	0.4842	0.1396	0.0345*	0.4562	1.0000	
Jun	0.0593	0.8078	0.7210	0.2904	0.2087	0.2697	0.0223*	0.3104	0.8314	1.0000

Note: * <0.05

Table 2 lists the overall incidence (% of total isolated colonies) of the identified taxa for the 2-year study and their occurrence frequency [O = occasional (1–5 months), F = frequent (6–8 months), or P = permanent (>8 months)]. Each taxon that has been specifically reported to cause disease is denoted by an asterisk. An overall 91 taxa from 31 zygomycetes, ascomycete, and deuteromycete genera were isolated.[26] The most frequently encountered genera were *Penicillium, Aspergillus, Alternaria, Phoma,* and *Fusarium,* · which, together with yeasts, accounted for 69% of all the colonies counted. The most frequently encountered identified species were *A. niger, A. fumigatus, Alt. alternata,* and *A. terreus. Penicillium* isolates were common but were not identified to species.

Although all fungi can potentially cause disease, 10,048 colonies (38% of the total colonies) were members of taxa (30 in all) that are commonly considered to cause disease (hypersensitivity or infectious) according to the literature (see Table 3). The monthly recoveries of these fungi (Figure 2) varied considerably from school to school. Table 4 shows the number of taxa and colonies, as well as the percentage of these taxa, isolated in each school. The most clearly differentiated schools were numbers 1, 3, and 9. School 1 had the highest total number of colonies, but the lowest percentage of common disease-related taxa (14%); a similar number of total colonies was recovered from school 9, yet the percentage of disease-related taxa reached 48%; and in school 3, 67% of the fungal taxa were those commonly considered to cause disease. Linear regression analysis indicated no correlation between total colonies and common disease agents at any sampling site.

The incidence of commonly reported disease agents most frequently encountered in the dust, that is *Alt. alternata, A. fumigatus, A. niger, A. terreus, Cladosporium cladosporioides, Paecilomyces variotii,* and *Penicillium* species (Figure 3), varied markedly throughout the year and from one year to the other. The Mann-Whitney test[27] was performed on data for each of these important taxa most frequently encountered in both public and private schools, but no significant difference was discovered for any taxon. The Chi-square test, when applied to the values shown in Table 5, yielded a value of 85.14, which corresponds to a slight association between type of school and taxon. This may be attributed to this test's high sensitivity when applied to a factor (the taxa) with so many levels, so it may be concluded that the pattern of taxa in private and state schools is barely differentiated.

V. DISCUSSION

Thirty-eight percent of the isolated colonies belonged to taxa recognized as important agents of disease. This percentage is similar to that reported by other authors.[2]

In broad terms, there were seasonal variations in concentrations of common disease agents isolated from dust (Figure 2). The highest total number

TABLE 2 Taxa Identified During the Years Studied

	Taxa	Incid	Pre		Taxa	Incid	Pre
1	Absidia ramosa	0.06	O	48	Eurotium chevalieri*	0.15	F
2	Acremonium alternatum	0.07	O	49	Fusarium spp.	1.79	P
3	Alternaria chartarum	0.09	O	50	Geotrichum candidum*	0.01	O
4	A. consortiale	1.37	F	51	Gliocladium spp.	0.01	O
5	A. dendritica	0.40	F	52	Monilia sitophila*	0.26	F
6	A. oleracea	0.21	F	53	Mucor circinelloides	0.11	F
7	A. sonchi	0.08	O	54	M. hiemalis	0.37	P
8	A. alternata*	2.24	P	55	M. plumbeus	0.04	O
9	A. tenuissima	0.44	P	56	M. racemosus	0.02	O
10	Alternaria spp.	0.25	F	57	Mucor spp.	0.11	O
11	Arthrinium phaeospermum	0.07	O	58	Mycotypha africana	0.01	O
12	Aspergillus alliaceus	0.02	O	59	M. microspora	0.02	O
13	A. candidus	0.19	F	60	Myrothecium roridum	0.02	O
14	A. carneus	0.18	F	61	M. striatisporum	0.01	O
15	A. eburneo-cremeus	0.06	O	62	M. verrucaria	0.01	O
16	A. flavipes	0.19	O	63	Nigrospora sphaerica	0.01	O
17	A. flavus*	0.02	O	64	Paecilomyces carneus	0.01	O
18	A. fumigatus*	4.36	P	65	P. farinosus	0.08	O
19	A. japonicus	0.08	O	66	P. variotii*	0.34	P
20	A. melleus	0.09	O	67	Paecilomyces spp.	0.01	O
21	A. niger*	8.32		68	Penicillium ser. Arenicola*	0.01	O
22	A. niveus	0.19	F	69	P. ser. Camembertii*	0.02	O
23	A. ochraceus*	0.03	O	70	P. ser. Citreonigra*	0.02	O
24	A. oryzae	1.22	P	71	P. ser. Citrina*	1.85	P
25	A. parasiticus*	0.06	O	72	P. ser. Expansa*	0.58	F
26	A. petrakii	0.09	O	73	P. ser. Fellutana*	0.01	O
27	A. silvaticus	0.11	O	74	P. ser. Glabra*	3.25	P
28	A. speluneus	0.10	O	75	P. ser. Implicata*	0.04	O
29	A. sydowi	0.39	F	76	P. ser. Islandica*	1.69	P
30	A. terreus*	2.01	P	77	P. ser. Miniolutea*	0.35	O
31	A. ustus	0.44	P	78	P. ser. Olsonii*	0.01	O
32	A. versicolor	0.81	P	79	P. ser. Oxalica*	0.10	O
33	Aspergillus spp.	0.23	P	80	P. ser. Restricta*	0.50	F
34	Aureobasidium pullulans*	0.01	O	81	P. ser. Urticicola*	5.91	P
35	Botrytis cinerea	0.01	O	82	Penicillium spp.*	5.05	P
36	Cladosporium cladosporioides*	0.86	F	83	Phoma spp.	4.82	P
37	C. macrocarpum	0.01	O	84	Rhizopus nigricans	0.91	P
38	C. sphaerospermum	0.05	O	85	Sartorya fumigata	0.01	O
39	Chaetomium sp.	0.05	O	86	Syncephalastrum racemosum	0.02	O
40	Cunninghamella echinulata	0.01	O	87	Torula spp.	0.02	O
41	Curvularia brachyspora	0.01	O	88	Trichoderma harzianum	0.30	F
42	C. lunata	0.12	F	89	T. koningii	0.29	F
43	C. oryzae	0.01	O	90	T. viride*	0.01	O
44	Drechslera australiensis	0.78	P	91	Ulocladium alternariae	0.01	O
45	D. biseptata	0.19	O	92	Mycelia Sterilia	17.04	P
46	D. ravenelii	0.01	O	93	Yeasts	18.77	P
47	Emericella nidulans*	0.08	O	94	Unknowns	8.35	P

Note: *, Pathogenic; Incid, incidence (% of total colonies); Pre, presence; O, occasional; F, frequent; P, permanent.

TABLE 3   Potential Pathogenicity to Man and Clinical Features of Some of the Isolated Taxa

Taxa	Potentially Induced Pathologies	References
Alternaria alternata	Allergies, mycotoxicoses, Baker's asthma	3, 8, 40–44
Aspergillus fumigatus	Allergic bronchopulmonary aspergillosis, alveolitis, sporosis, opportunistic infections	3, 8, 45–51
A. niger	Allergic bronchopulmonary aspergillosis, sporosis, opportunistic infections	3, 42, 44, 51
A. terreus	Allergic bronchopulmonary aspergillosis	51
Cladosporium cladosporioides	Allergies, pulmonary mycosis	44, 52
Paecilomyces variotii	Animal paecilomycosis and mycotoxicosis	21
Penicillium spp.	Alveolitis, pneumonia, sporosis, mycetoma, asthma, surface infections	3–4, 8, 42, 52–63

of colonies was recorded in September. The Wilcoxon test showed a significant difference between September and all other months except May and June (dust was not collected during July or August). This agrees with the findings of Tuñon de Lara et al.,[28] who noted the greatest concentration of fungi in summer, suggesting that other factors besides temperature and humidity must influence the development of dust fungi. On comparing the results obtained in the present study and others previously reported by our group,[29–34] it is apparent that the concentration peaks of some taxa (viz. *Alt. alternata*, *A. niger*, and *C. cladosporioides*) are reached in dust at a later date than they are in air. This observation also agrees with the above-mentioned authors,[28] who found no correlation between concentrations of airborne and dust fungi.

Table 3 shows pathological processes which may be induced by some of the isolated fungi. Of the species belonging to the genus *Alternaria* which were isolated in the course of this study, the consulted literature listed only *Alt. alternata* as an allergen. This species, found every month at all sampling sites, accounted for 2.24% of total isolates. The materials on which it usually develops (e.g., dead leaves and other organic material) may explain its notable presence in school 12 which is situated in a semi-rural area bordering on farming land and abundant vegetation.

Seven of the 21 isolated species of the genus *Aspergillus* are described as recognized disease agents. Several authors[16–18,22,28] have described *Aspergillus* as one of the common groups of fungi in domestic dust, perhaps due to the ease with which it colonizes the material contained therein. Notable numbers of colonies of the following species were recovered: *A. fumigatus*, which comprised 4.36% of total isolates and was always present in dust, being particularly prominent in school 10; *A. niger*, the most abundant *Aspergillus* species (8.32% of total isolates), was also permanently present in dust and was present in especially high levels in school 3; and *A. terreus* (2%), which was present perennially and which reached the highest levels in school 12.

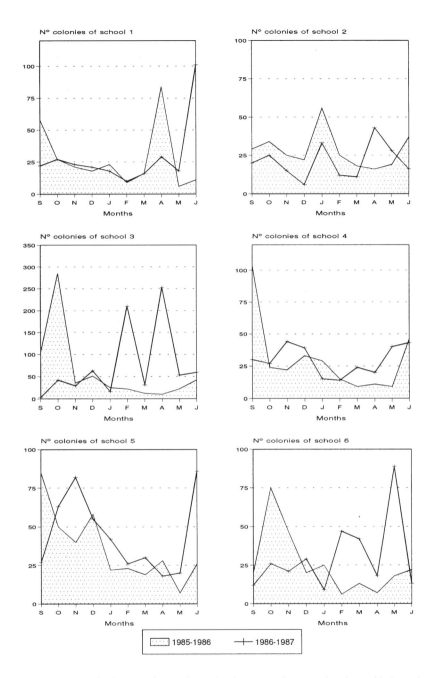

FIGURE 2 Graphs showing the numbers of pathogenic colonies isolated monthly in each school during the two years studied.

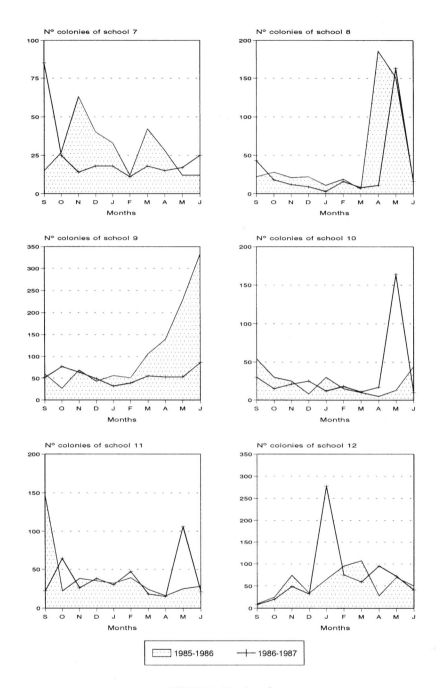

FIGURE 2 *(Continued)*

TABLE 4   Number of Taxa and Colonies

	Schools	No. Taxa		No. Colonies	
		Total	Pathogenic	Total	Pathogenic
1	Córdoba	52	19	3924	558 (14%)
2	San Juan de la Cruz	67	22	1743	490 (28%)
3	Julio Romero de Torres	63	21	2041	1367 (67%)
4	Cervantes	64	21	2593	616 (24%)
5	Santuario	63	19	2026	807 (40%)
6	Averroes	56	22	1753	559 (32%)
7	San Rafael	51	16	1287	530 (41%)
8	Enríquez Barrios	53	19	1515	782 (52%)
9	La Milagrosa	59	20	3478	1673 (48%)
10	San Miguel	52	20	1271	557 (44%)
11	José de la Torre y del Cerro	58	18	2067	793 (38%)
12	Ahlzahir	58	18	2640	1316 (50%)

In previous works,[30,31] we stated the quantitative importance of the species of *Cladosporium* in air. However, as can be seen from the results of this chapter, their relative abundance in dust is quite low. This finding has been reported by several authors.[16,35–39] The harsh conditions often found in dust,[35] including low water availability, may provide less than optimal conditions for this genus. *C. cladosporioides*, a recognized allergenic species, was, however, frequently present in dust. *Paecilomyces variotii* was found in every dust sample, contributing 0.34% to the total. *Penicillium*, a cosmopolitan genus that commonly occurs indoors, was the most abundant genus (19.4% of all isolated colonies), and has also been reported by other authors as dominant in schools[17,18,22] and in domestic dust.[16,24,35–37]

Linear regression analysis revealed no association at any sampling site between the total number of colonies and the number of common disease agents. We believe that the characteristics of the school must somehow influence the types of fungi that grow there and are in agreement with Tuñon de Lara et al.[28] who report that fungal concentrations in dust are higher in old buildings, the environments of which may play an important role in fungal development. Two of the buildings which recorded markedly different results from the others may serve to illustrate this point. School 9 is more than a century old, and the other, school 3, is partly composed of an old building. School 1 also displayed a prominent spore count, possibly due to its proximity to the River Guadalquivir.

It is worth noting that the mean cfu level per gram of school dust, 2888, was much lower than that reported by other authors for house dust,[16,36,37] a difference that may be attributable to the obvious difference between the school environment and that of the home, but that could also be related to differences in sample collection, handling, and analytical methodology. The relatively

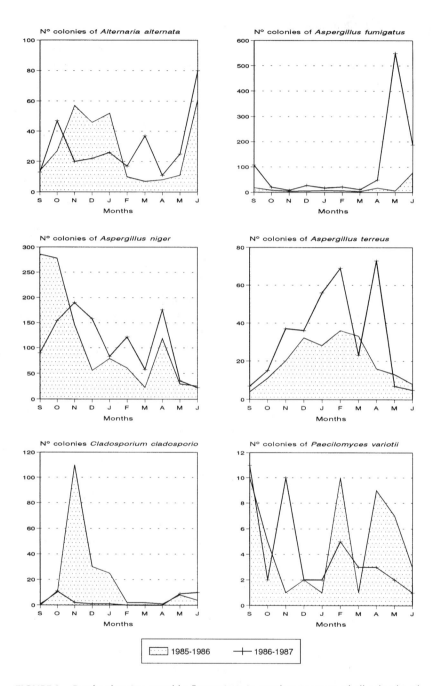

FIGURE 3 Graphs showing monthly fluctuations in number (average of all schools) of frequently encountered pathogenic fungi recovered during the two years studied.

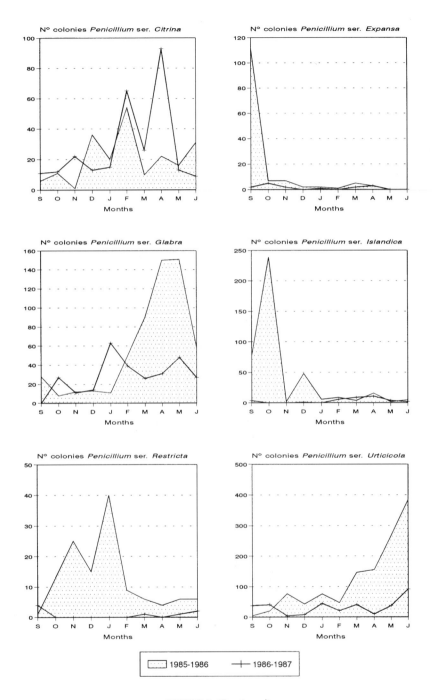

FIGURE 3 *(Continued)*

TABLE 5 Chi-Squared Analysis on Number of Colonies in Schools

Schools	No. of Total Colonies of the Most Abundant Pathogenic Taxa								
	sp1	sp2	sp3	sp4	sp5	sp6	sp7	sp8	Total
Private	322	548	1002	292	141	39	36	2829	5209
Public	269	601	1190	237	86	30	54	2278	4745
Total	591	1149	2192	529	227	69	90	5107	9954

Note: Chi-square = 85.14. D.F. = 7. Significance = 99.9%. sp1 = *Alternaria alternata*; sp2 = *Aspergillus fumigatus*; sp3 = *Aspergillus niger*; sp4 = *Aspergillus terreus*; sp5 = *Cladosporium cladosporioides*; sp6 = *Monilia sitophila*; sp7 = *Paecilomyces variotii*; sp8 = *Penicillium* spp.

short duration of the study — two years — has not produced sufficient information to permit a definitive statement with respect to patterns of taxon prevalence in schools. Nevertheless, the guidelines for the reduction of fungi remain the same as for any other environment: schools should be kept meticulously clean (to prevent dust from accumulating in cracks and corners) and adequate ventilation must be provided.

ACKNOWLEDGMENTS

The authors are grateful to the Spanish DGICyT for financial support granted through Project SM 89-0010 of the Sanitary Research Fund.

REFERENCES

1. Caplin, I. and Haynes, J.T., Mold allergy, *Ann. Allergy*, 28, 87, 1970.
2. Bronswijk, J.E.M.H., Rijckaert, G., and Lustgraaf, B., Indoor fungi, distribution and allergenicity, *Act. Bot. Neel.*, 35(3), 329, 1986.
3. Hyde, H.A., Atmospheric pollen and spores in relation to allergy, *Clin. Allergy*, 2, 153, 1972.
4. Calvo, M.A., Guarro, J., and Suarez, G., Los hongos como agentes etiológicos de alergias y enfermedades pulmonares: su incidencia en Barcelona, *An. Med. Cirug.*, 56, 329, 1976.
5. Prince, H.E. and Meyer, G.H., An up-to-date look at mold allergy, *Ann. Allergy*, 37, 18, 1976.
6. Al-doory, Y., Domson, J.F., Howard, W.A., and Michael, R., Airborne fungi and pollens of the Washington D.C., Metropolitan area, *Ann. Allergy*, 45, 360, 1980.
7. Maunsell, K., The impact of aerobiology on allergy, *Act. Allergol.*, 26, 329, 1971.
8. Gravesen, S., Fungi as a cause of allergic disease, *Allergy*, 34, 135, 1979.
9. Salvaggio, J. and Aukrust, L., Mold-induced asthma, *J. Allergy, Clin. Immunol.*, 68, 327, 1981.
10. Petersen, B.N. and Sandberg, I., Diagnosing allergic diseases by correlating pollen/fungal spore counts with patient scores of symptoms, *Grana*, 20, 219, 1981.
11. Berrens, L. and Young, E., Purification and properties of house dust allergens, *Int. Arch. Allergy*, 19, 341, 1961.
12. Voorhorst, R., Spieksma-Boezeman, M.I.A., and Spieksma, F.T.M., Is a mite (*Dermatophagoides* sp) the producer of the house-dust allergen?, *Allergy Asthma*, 10, 329, 1964.
13. Berrens, L., *The Chemistry of Atopic Allergens*, Vol. 7, Monographs in Allergy, S. Karger, Basel, 1971, 135.
14. Olive, A., Alergenos del polvo doméstico. I. Estudio del papel de las diferentes fracciones del polvo doméstico, *Allergol. Immunopathol.*, 3, 137, 1975.
15. Collins-Williams, C., Hung, F., and Bremner, K., House dust mite and house dust allergy, *Ann. Allergy*, 37, 12, 1976.
16. Horak, B., Preliminary study on the concentration and species composition of bacteria, fungi and mites in samples of house dust from Silesia (Poland), *Allergol. Immunopathol.*, 15(3), 161, 1987.
17. Gravesen, S., Larsen L., and Skov P., Aerobiology of schools and public institutions — part of a study, *Ecol. Dis*, 2(4), 411, 1983.
18. Gravesen, S., Larsen L., Gyntelberg F., and Skov P., Demonstration of microorganisms and dust in schools and offices, *Allergy*, 41, 520, 1986.

19. Sederberg-Olsen, J. and Hardt, F., The indoor climate in schools, *Ugeskr. Laeger*, 146, 1453, 1984.

20. Nexo, E., Skow, P., and Gravesen, S., Extreme fatigue and malaise: a syndrome caused by malcleaned wall-to-wall carpets?, *Ecol. Dis.*, 2, 415, 1985.

21. Samson, R.A., Occurrence of moulds in modern living and working environments, *Eur. J. Epidemiol.*, 1(1), 54, 1985.

22. Gravesen, S., Microbial and dust-pollution in non industrial work places, *Adv. Aerobiol. EXS.*, 51, 279, 1987.

23. WHO (World Health Organization), Indoor air pollutants: exposure and health effects, *WHO-Meeting in Nördlingen* 1982, Eur. Report No. 78, Copenhagen, 1983.

24. Macher, J.M., Inquiries received by the California indoor air quality program on biological contaminants in buildings, *Adv. Aerobiol. EXS.*, 51, 275, 1987.

25. Tuite, J., *Plant pathological methods fungi and bacteria*, Burgess Publishers, San Francisco, 1969.

26. Alexopoulus, C.J. and Mims, C.W., *Introducción a la Micología*, Omega, Barcelona, 1985.

27. Caridad Ocerín, J.M., *Analisis de datos con B.M.D.P.*, Servicio de publicaciones, Universidad de Córdoba, 1989.

28. Tuñon de Lara, J.M., Tessier, J.F., Lafont-Grellety, J., Domblides, P., Mary, J., Faugère, J.G., and Taytard, A., Indoor moulds in asthmatic patients homes, *Aerobiologia*, 6, 98, 1990.

29. Infante, F., *Identificación, cuantificación y variación estacional de microhongos aerovagantes de interior y exterior en hogares de la ciudad de Córdoba*, Doctoral Thesis, Fac. Ciencias, University of Córdoba, 1987.

30. Infante, F. and Dominguez, E., Incidencia de esporas de *Cladosporium* Link ex Fr. en los habitats domésticos de la ciudad de Córdoba, *Act. VI Simp. Nac. Bot. Cript.*, Granada, 281, 1987.

31. Infante, F. and Dominguez, E., Annual variation of *Cladosporium* spores in homes habitats in Córdoba, Spain, *Ann. Allergy*, 60(3), 256, 1988.

32. Infante, F., Dominguez, E., Ruiz De Clavijo, E., and Galan, C., Incidence of *Alternaria* Nees ex Fr. in dwelling of Córdoba city (Spain), *Allergol. Immunopathol.*, 15(4), 221, 1987a.

33. Infante, F., Ruiz De Clavijo, E., Galan, C., and Gallego, G., Ocurrence of *Alternaria* Nees ex Fr. in indoor and outdoor habitats in Córdoba (Spain), *Adv. Aerobiol. EXS.*, 51, 157, 1987b.

34. Infante, F., Ruiz De Clavijo, E., Galan, C., and Dominguez, E., Estudio comparativo de *Alternaria* Nees ex Fr. en el aire de exterior e interior en la ciudad de Córdoba, *An. Assoc. Palinol. Leng. Esp.*, 3, 5, 1987c.

35. Davies, R.R., Viable moulds in house dust, *Trans. Brit. Mycol. Soc.*, 43(4), 617, 1960.

36. Ales, J.M., Canto, G., Garcia, L.M., Jimenez-Diaz, C., Lahoz, F., Ortiz, F., and Sastre, A., Papel etiológico de los hongos del aire en el asma bronquial, *Rev. Clin. Esp.*, 64(3), 143, 1957.

37. Jimenez-Diaz, C., Ales, J.M., Ortiz, F., Lahoz, F., Garcia-Puente, L.M., and Canto, G., The aetiologic role of molds in bronchial asthma, *Act. Allergol.*, 15, (Suppl. 7), 139, 1960.

38. Dransfield, M., The fungal air-spora at Samaru Northern Nigeria, *Trans. Br. Mycol. Soc.*, 49, 121, 1966.

39. Gravesen, S., Identification and prevalence of culturable mesophilic microfungi in house-dust from 100 Danish Homes, *Allergy*, 33, 268, 1978.

40. Ciegler, A., Fungi that produce mycotoxins: conditions and occurrence, *Mycopathology*, 65, 5, 1979.

41. Bonilla-Soto, D., Rose, N.R., and Arbesman, C.E., Allergenic molds: antigenic and allergenic properties of *Alternaria alternata*, *J. Allergy*, 32, 246, 1961.

42. D'Amato, G., Cocco, G., Ruggiero, G., and Sales, L., Importanza dei miceti nella patologia respiratoria su base allergica. I. Incidenza di positivita ai test cutanei, *Arch. Monaldi Tisiol. Malat. Apparat. Respir.*, 30, 1, 1975.

43. Muñoz, F. and Martin, M.A., *Alergia a hongos*, Sandoz, Barcelona, 1983.
44. Bronswijk, J.E.M.H. and Sinha, R.N., Role of fungi in the survival of *Dermatophagoides* (Acarina: *Pyroglyphidae*) in house-dust enviroment, *Environ. Entomol.*, 2, 142, 1973.
45. Gregory, P.H., *The Microbiology of the Atmosphere*, 2nd ed., Leonard Hill, London, 1973.
46. Mullins, J., Harvey, R., and Seaton, A., Sources and incidence of airborne *Aspergillus fumigatus* (Fres.), *Clin. Allergy,* 6, 209, 1976.
47. Calvo, M.A., Guarro, J., and Vicente, E., Presencia de *Aspergillus fumigatus* en la atmósfera urbana, *Anal. Med. Cir.*, 251, 69, 1978.
48. Frankland, A.W. and Davies, R.R., Allergie aux spores de moisissures en Angleterre, *Le Poumon et le Coeur*, 1, 11, 1965.
49. Turner, K., Elder, J., O'Mahony, J., and Johansson, S., The association of lung shadowing with hypersensitivity responses in patients with allergic broncho-pulmonary aspergillosis, *Clin. Allergy,* 4, 149, 1974.
50. Dessaint, J., Bout, D., Fruit, J., and Capron, A., Serum concentration of specific IgE antibody against *Aspergillus fumigatus* and identification of the fungal allergen, *Clin. Immunol. Immunopathol.*, 5, 314, 1976.
51. Sandhu, R.S., Mehta, S.K., Khan, Z.U., and Singh, M.M., Role of *Aspergillus* and *Candida* species in allergic bronchopulmonary mycoses a comparative study. *Mycopathology,* 63, 21, 1978.
52. Chirila, M., Capetti, E., and Banescu, O., The relationship between air-borne fungal spores and *Dermatophagoides pteronyssinus* in the house dust, *Rev. Roum. Med. Int.*, 19, 73, 1981.
53. Bessot, J.C. and Pauli, G., Prevèntion de l'allergie respiratoire aux acariens de la poussière de maison, *Rev. Fr. Allergol.*, 25(3), 155, 1985.
54. Canto, G. and Jimenez-Diaz, C., Estudio de los hongos en el aire de Madrid durante un año, *Rev. Clin. Esp.*, 17(4), 226, 1945.
55. Frouchtman, R., Contribución al estudio de las alergopatías respiratorias climáticas en Barcelona. Importancia de las bacterias del aire, *Rev. Clin. Esp.*, 23(4), 292, 1946.
56. Alemany-Vall, R., Sensibilidad respiratoria a hongos, *Med. Clin.*, 13, 102, 1949.
57. Aller, B., Rey, M., and Martinez, A., Estudio de la incidencia de los hongos de León durante un año, *Rev. Clin. Esp.*, 121(5), 13, 1971.
58. Lacey, J., The air spora of a portuguese cork factory, *Ann. Occup. Hyg.*, 16, 223, 1973.
59. Faraco, B.F.C. and Faraco, B.A., Mycological pollution of the atmosphere, *Rev. Bra. Med.*, 31(11), 779, 1974.
60. Requejo, V.M., Micoflora atmosférica de la ciudad de Trujillo (Perú). III. Géneros aislados durante el año 1971, *Mycopathology*, 56(1), 15, 1975.
61. Gravesen, S., On the connection between the occurrence of airborne microfungi and allergy symptoms, *Grana*, 20, 225, 1981.
62. Solley, G.O. and Hyatt, R.E., Hypersensitivity pneumonitis induced by *Penicillium* species, *J. Allergy Clin. Immnunol.*, 65, 65, 1980.
63. Larsen, L.S., A three-year survey of microfungi in the outdoor air of Copenhagen 1977–1979, *Grana*, 20, 197, 1981.

Chapter 6

PREVALENCE OF HOUSE DUST MITES FROM HOMES IN THE SONORAN DESERT, ARIZONA

Mary Kay O'Rourke
Cathy Lee Moore
Larry G. Arlian

CONTENTS

I. ABSTRACT

House dust mites are rarely found in homes with low indoor relative humidity (<35%). Tucson is an urban region in the arid Sonoran Desert with frequent outdoor relative humidities of less than 20%. A previous study evaluated skin test reactivity within the general population and among those with skin test reactivity to screening antigens. Among atopics, only 3.5% reacted to *Dermatophagoides farinae* with a wheal ≥5 mm. Based on these facts, we hypothesized that few Tucson homes support populations of house dust mites. To test the hypothesis, we collected dust from four locations (mattress, bedroom floor, couch, and living room floor) in 82 randomly selected homes. *D. farinae* were found in 67% of the homes sampled; *D. pteronyssinus* were found in 8.5% of the homes containing mites. No mites were collected in 31% of the homes. In most temperate regions, the greatest number of mites are found during the late summer and early fall. In Tucson, we collected the greatest number of mites in winter (2116 mites/g of fine dust; composite value of four locations from one house). Live mites, gravid females, and eggs were most commonly collected under conditions of high humidity. Subjects previously exhibiting skin test reactivity to other antigens (n = 134) were tested with *D. farinae* extract. Of these, 12.6% responded with a wheal 5 mm greater than the negative control, suggesting that house dust mites may be a greater allergy problem in Tucson than previously thought. These data are preliminary, representing one collection year. We reject the initial hypothesis since homes with mites were commonly observed in Tucson. We propose that elevated indoor humidity caused by evaporative cooler use may sustain mites during the driest and hottest part of the year (May–June).

II. INTRODUCTION

The house dust mites, *Dermatophagoides farinae, D. pteronyssinus,* and *Euroglyphus maynei,* live on skin flakes accumulating in upholstered furniture, carpets, and bedding. Their bodies, feces, and other secretions are sources of common allergens.[1-4] With exposure exceeding threshold levels, genetically susceptible people develop allergies and sometimes asthma.

Mites require specific temperature and humidity regimes to thrive.[5-7] Mite populations expand when indoor relative humidity (rh) is 60–75% or above at temperatures of 60–75°F.[8,9] Based on laboratory results, the critical equilibrium humidity for fasting *D. pteronyssinus* is 73% rh at 25°C,[10] and 55–75% rh proportional to temperatures of 15–35°C for *D. farinae*.[11] Most mites die in 1-3 days at 40–50% rh; but some mites survive for 4–8 days.[12,13] Since mites require relatively high humidities for survival, most surveys find the greatest mite prevalence in humid, temperate regions and the least in arid regions.

Arid climatic conditions are unfavorable for the development of mite infestations in homes. For example, 12 of 20 homes sampled in Denver, Colorado, lacked mites.[14] An additional 20% contained mite densities of <40 mites/g of dust. The average daily rh in Denver is <40–50% with little seasonal variation. However, 4 of the 20 homes contained mite concentrations ranging from 100–360 mites/g of dust, including live mites. Microclimatic conditions in these homes were apparently favorable for mite growth despite the regional aridity.

Tucson, an urban area with more than 600,000 people, is located in the arid Sonoran Desert of the southwestern United States. In a previous Tucson study among skin test responsive subjects screened from the general population, 26.5% had some skin test reactivity to *D. farinae*, but only 3.5% exhibited skin test wheals ≥5 mm.[15] Further, for all 14 skin test antigens used, response varied with age.[16] Subjects responding to mites in the cited studies had usually migrated from temperate regions. Yet, there was no evidence to confirm or deny exposure to dust mites in Tucson homes. Based on the low skin test reactivity ≥5 mm, and the low outdoor humidity (frequently <20%), we hypothesized that mites would be absent from house dust samples collected in Tucson. This survey was undertaken, in part, to test this hypothesis by determining the prevalence and concentration of house dust mites in Tucson homes.

III. METHODS

In Tucson, Pima County Government employs more than 4500 people. We surveyed all county employees and enrolled 2322 in a long-term study addressing respiratory health. The recruited population reflects the same age, gender, ethnicity, smoking behavior, and respiratory disease rates reported for the working population of Pima County.[17] For this study, 82 homes were randomly selected from the enrolled study population. All subjects had lived in Tucson for more than five years. We characterized the living environment and assessed subject health using the Standard Environmental Inventory Questionnaire,[18] a version of the Standard Health History and Status Questionnaire,[19] and a basic health workup that included spirometry, a blood draw, and 27 scratch skin tests including positive and negative controls.[15,16,20] In about half the cases, subjects received questionnaires and medical tests in their homes from trained personnel; otherwise, subjects were evaluated in the clinic.

A. DUST COLLECTION

We collected dust samples from four locations in each home. The locations were: the mattress in the master bedroom, the bedroom floor, the most frequently occupied piece of furniture in the main room, and the main room floor adjacent to the sampled furniture. The sites sampled and the methodology employed comply with the recommendations of the WHO (World Health Organization) working group on house dust mites and asthma.[21,22] An area of 1 m² was vacuumed for two minutes. These samples were collected using a 2.2 hp Hoover Port-a-Power™ vacuum cleaner. Dust was deposited on a preweighed bed sheet filter 625 cm² in area and 180 threads per inch. Filters were held in a wand attachment dust trap used for vacuum demonstrations. Filters were removed from the dust trap and secured with a rubber band.

In the laboratory, the rubber band was removed, the filters were reweighed and the total dust weight was calculated. Collected house dust was deposited stratigraphically in two layers on the filter. The upper layer was a fibrous mat composed of hair, fiber, lint, and other relatively low density particles. The upper layer was moved aside easily with a laboratory spatula. The lower layer, composed of comparatively fine-grained particulate, was analyzed for mites.

B. EXTRACTION AND IDENTIFICATION

Fifty milligrams of the fine house dust stratum were placed in a 100-ml beaker with a drop of liquid soap (surfactant) and 30 ml of saturated NaCl solution. The beaker contents were mixed thoroughly using a vibrating mixer for 30–40 sec. The mixture was poured through a 45-μm sieve. Several drops of crystal violet stain were added to the residue and rinsed through the sieve with low speed tap water. The residual sediment was transferred to a gridded Petri dish and examined for mites. Mites were removed, cleared with lactic acid, and identified using a compound microscope at 400×. Mite concentrations were calculated and expressed as mites/g of dust.

C. QUALITY ASSURANCE

All samples were initially analyzed at the University of Arizona (UA), Tucson. A stratified subset of samples (~10% of the total) based on initial mite concentration (UA) was sent to Wright State (WS), Dayton, Ohio, for reanalysis. This subset was selected as follows: one third appeared mite free; one third contained moderate mite concentrations; and one third contained large numbers of mites. Samples were considered similar if numbers varied by ≤50%.

D. SKIN TESTING

In homes that were sampled for house dust mites, occupants were asked if they would be willing to be tested for skin test reactivity to a variety of allergens. Consenting subjects were skin tested with *D. farine* (30,000 AU/ml) and five screening antigens: house dust (1:10), *Alternaria* spp. (1:100), *Cynodon dactylon* (1:20), tree mix (1:20), and weed mix (1:20), positive and negative controls. All extracts were provided by Bayer Laboratories (Spokane, WA). Scratch tests were performed according to procedures described by

Pepys.[20] After 20 min the skin test results were recorded. A zero value was recorded for all test sites where no wheal developed. When a wheal was observed, the diameter was measured in two directions at right angles. The two measures were summed and the sum recorded in millimeters. Occurrence of pseudopods was noted. Subjects were deemed atopic (that is, capable of forming IgE to common allergens) if the sum of the wheals from the five screening antigens ≥8 mm.

IV. RESULTS

Eighty-two homes were evaluated for house dust mite prevalence. Of the 303 subjects living in these homes, 259 agreed to be evaluated for skin test reactivity. Using the criterion described above, 134 of the 259 subjects screened were classified as atopic. Skin test responses of <5 mm and ≥5 mm to *D. farinae* were found among 7.2% and 12.6% of the atopic participants, respectively. No wheal developed to *D. farinae* in 80.2% of those deemed atopic. One person lacking other skin test responses reacted to *D. farinae*. Results were analyzed in this fashion for comparison with those of Brown et al.[15] and Barbee et al.[16] We examined house dust samples for *D. farinae, D. pteronyssinus, E. maynei,* and *Blomia tropicalis.* Low concentrations (<100 mites/g) of *D. pteronyssinus* were found in 8.5% of the homes containing mites (or 3.5% of all houses sampled) and accounted for less than 1% of all mites collected. Ninety-nine percent of all mites collected in Tucson were *D. farinae*. Neither *E. maynei* nor *B. tropicalis* were found in any of the samples collected.

Table 1 shows the percentage of homes with mites. Values are presented by location (i.e., mattress, main room carpet, etc.) and by mite concentration (i.e., ≥100 mites/g) for all houses sampled (n = 82) and for the mite positive homes only (n = 56). Two thirds of the houses contained house dust mites.

TABLE 1   Percent of Houses with Any Mites and Percent of Houses with More Than 100 Mites/g House Dust by Location

	Total Houses Sampled (n=82)		Mite Positive Houses (n=56)
Locations	% of Houses with any Mites	% of Houses with Mite Conc. ≥100 Mites/g House Dust	% of Mite Positive Houses with Conc. ≥100 Mites/g House Dust
Furniture (main room)	26.8%	0.5%	18.2%
Carpet (main room)	34.1%	12.2%	35.7%
Mattress (bedroom)	32.5%	14.3%	48.0%
Carpet (bedroom)	42.1%	20.7%	45.9%
Household composite sample (4 locations/ house)	67.1%	17.1%	25.4%

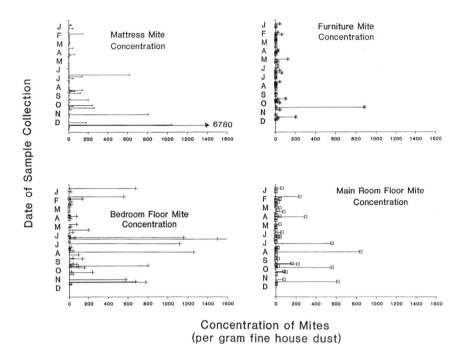

Concentration of Mites
(per gram fine house dust)

FIGURE 1 *Dermatophagoides farinae* seasonal concentrations from collection sites within sampled homes.

Carpet samples often had more mites than did furniture in the same room; generally bedrooms had more mites than the main rooms. Figure 1 presents site-specific mite concentrations for each site in each home arranged by season of sampling.

Figure 2 illustrates composite household mite concentrations. These values were derived by summing the mite counts from the four locations within a house and dividing by the combined weight of the subsamples. Mite concentrations were consistently low in homes sampled from February through mid-June. The greatest concentrations were found in homes sampled during the late summer and winter. Concentrations in fall-sampled homes were marginally lower than in those sampled in late summer. Figure 3 summarizes the data in Figure 2 by season. More mites were recovered from homes sampled in summer and winter seasons as opposed to those sampled during the spring and fall.

Figure 4 presents the growth mite stages collected seasonally. The lower portion of the bar (connected by the lower background line) represents the concentration of mature mites. The lower two courses describe the concentration of male vs. female mites collected. The concentration of mature mites parallels the total mite concentration (the upper background line). The quiescent protonymph stage was recovered during all seasons, but in homes sampled

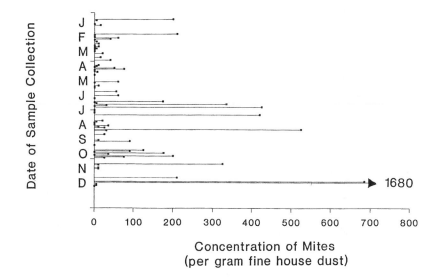

FIGURE 2   Composite house dust mite concentrations; results are combined from the four subsamples presented in Figure 1.

during the late spring where mite concentrations were low, protonymph concentrations equaled those of the summer.

Forty site-specific dust samples analyzed at UA, were sent to WS for quality assurance checks. One sample had too little dust for analysis. In the remaining 39 samples, neither group found mites in seven samples, and both found similar concentrations in 25 samples. In each of the remaining samples (n = 7), one or two mites were found by one group and not by the other. (UA found mites in three samples not reported by WS; WS found mites in four samples not reported by UA). Among the 39 QA samples, 99% of all mites identified were *D. farinae*. In the rare instances of *D. pteronyssinus* presence, the species was represented by one or two mites. Among all samples, UA reported two occurrences of *D. pteronyssinus* and WS reported three. These reports were mutually exclusive. UA sent three slides with *D. pteronyssinus* to WS; the identifications were confirmed.

V. DISCUSSION

An earlier study of skin test reactivity among 311 Tucson residents found that only 3.5% of the 157 atopic subjects responded to *D. farinae*.[15] The current

Total Mite Concentration
Seasonal Mean

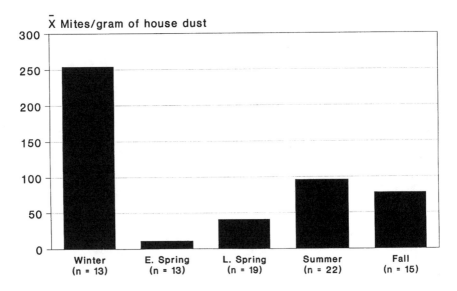

n = the number of homes sampled

FIGURE 3   Seasonal mite concentrations were derived by averaging values in Figure 2. Although the winter values are inflated by one apparently high value, mite concentrations tend toward greater numbers from late summer through fall and winter (August through January).

study (skin test total n = 259, atopics 134) shows a three- to fourfold increase in the number of atopics with a significant skin test response to *D. farinae* (12.6%). These are statistically significant differences (p =.05). Several factors might account for these different rates:

1. Too few people were assessed to adequately evaluate individual differences.
2. Although the extract has always been provided by the same company, there may be differences among batches. The original study was done prior to the development of a standardized extract. This study was done with the recently developed standardized extract.
3. House dust mite populations may have increased, resulting in greater subject exposure and skin test response.

Improved skin test extracts and limited population size are the most likely explanations. We have no mechanism of testing option 3.

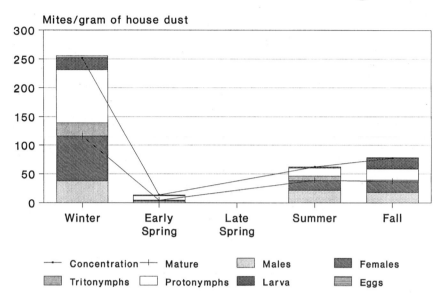

FIGURE 4 Mite growth stages represented by season.

D. pteronyssinus has an extensive range in Europe and was referred to as the "European house dust mite." By contrast, *D. farinae* was thought to be distributed solely in the Americas and became known as the "American house dust mite."[23] Recent work by Arlian et al.[24] describes mite concentrations in eight cities across the United States. Mites were found in all homes and more than 75% of the homes contained both species. The study clearly demonstrates that these two mite species coexist in the United States and the common names are misnomers.

In Tucson, *D. farinae* was the dominant mite collected in 97% of the mite-positive homes. *D. pteronyssinus* was rarely collected (only 8.5% of positive homes). These results were consistent with the resistance to relatively low humidity reported for *D. farinae*.[5,8,9] Mite survival in Tucson may be facilitated by the presence of a quiescent protonymph described by Wharton;[23] the development of the quiescent protonymph enables survival during prolonged drought.[25]

In samples examined in eight cities, Arlian et al.[24] found mites in every home. In Tucson, we appear to have mites in fewer homes and at lower concentrations. These differences may be caused by the criteria used for household selection. Our Tucson study evaluated randomly selected homes. Arlian et al.[24] examined 252 homes of house dust mite sensitive asthmatics. The development of skin test sensitivity requires exposure to mites, thus increasing the probability of (1) finding mites in a home, and (2) finding mites in elevated numbers. Our study was devised to characterize the mite prevalence in Tucson regardless of subject health status.

Korsgaard[26] found that concentrations of 100 mites/g of house dust increased the relative risk of asthma development by sevenfold. Platts-Mills et al.[27] found that results of mite antigen assays and mite counts are strongly correlated. They relate 2 µg of antigen to 100 mites/g of house dust and 10 µg of antigen to 500 mites/g. In general, elevated mite and antigen concentrations are important risk factors for mite sensitive asthmatics.[27]

Arlian et al.[24] reported that in 75-100% of homes sampled, depending on city, mite concentrations were ≥100 mites/g dust. In Tucson, only 17.1% of all homes sampled (25.4% of the mite-positive homes) had composite mite concentration values exceeding 100 mites/g. However, too few homes have been sampled to definitively state that Tucson has a lower rate of house dust mite occurrence.

Figures 1 through 4 illustrate consistent mite presence in Tucson homes requiring rejection of the initial hypothesis. Mattress and furniture collections showed the highest mite concentrations. Seasonal trends appear site-specific during the late fall and winter months. Main room and bedroom floor concentrations were highest in homes sampled in summer, fall, and winter months. Presently, no interpretation can be made without a clear understanding of regional weather and evaporative cooler use patterns.

In other regions, researchers have interpreted house dust mite prevalence in response to seasonal weather patterns and house dust mite humidity and temperature requirements.[8,25,30] The same principles were applied to mite prevalence in Tucson. In his classic work, Seller[29] briefly describes Arizona's highly variable climate. Temperature patterns are similar to those of other temperate regions, but the magnitude of change differs. During winter, temperatures rarely descend below freezing and outdoor relative humidity is elevated. Daytime conditions are analogous to those typically encountered in late September–October in the Northeast; a time when house dust mites flourish. When frontal storms sweep into Tucson from the Pacific Northwest, Tucson is subjected to the "winter rainy season." Early spring has highly variable conditions but can be quite pleasant (i.e., May in the Northeast). Late spring is exceedingly dry. Relative humidity values of 5–10% are common, and daily high temperatures usually exceed 90°F and frequently exceed 100°F. The high temperatures stimulate the use of evaporative coolers, elevating indoor humidity and lowering indoor temperatures (about 20°F below outdoor temperatures).

Evaporative cooler use begins in late April or early May, so that, although this season has the lowest outdoor humidity in terms of regional outdoor weather, indoor humidities are elevated. Unlike air conditioners (refrigeration), evaporative coolers work by evaporating moisture from saturated cooler pads. The act of evaporation lowers the temperature of the intake air. The cooled, humidified air is pumped into the house and displaces the hot dry indoor air which is forced out through open doors or windows in another part of the house. The coolest wettest air sinks to the floor.

By early July, tropical moisture reaches Arizona from the south. As the valley floors heat up in the early part of the day, the moist air rises and clouds form. Hot summer mornings are followed by convective, monsoonal, afternoon thunderstorms. In the summer, homes receive moisture from both the regional humidity increase and the evaporative coolers. Summer rains usually end in late August or early September. In the fall, outdoor humidity declines rapidly although daytime temperature remains high and coolers are still used. Cooler use usually ends in mid-October. A detailed description of these seasons is found in O'Rourke.[31]

We believe that mite presence in Tucson depends primarily on the use of evaporative coolers. Once weather and cooler use patterns are understood, seasonal mite prevalence is readily explainable (Figures 2 to 4). In winter, regional humidity and temperatures are ideal for mite survival and reproduction. In early spring, the winter rains cease, humidities fall, temperatures increase, and mite survival and reproduction is low. In late spring, temperatures increase, evaporative cooler use begins, and indoor humidity increases primarily through the addition of cooled, humidified air from the evaporative cooler. Without the evaporative cooler, neither indoor temperature nor indoor humidity is likely to be suitable for mite reproduction. In our study, we began recovering mites from homes sampled in late spring when cooler use was intermittent throughout the day. Mite concentrations were elevated in homes sampled during the summer when indoor humidity was sustained at a high level through constant cooler use. In fall, outdoor humidity and temperatures declined and cooler use became intermittent and later ceased. We recovered slightly fewer mites in homes sampled during the fall.

Microhabitat biases were evident among samples collected within a home (Figure 1). Mite collections from mattresses and furniture were consistently low except in those homes sampled in winter. Furniture and mattresses were cooled, warmed, and humidified by circulating air and occupants. By contrast, mite concentrations from the carpets of both living room and bedroom were highest in homes sampled during high humidity conditions, and concentrations reflected the seasonal conditions recounted above. In Tucson, most buildings are concrete slab construction, with carpet laid over concrete.[28] Moisture may condense at the carpet–concrete interface and result in the development of appropriate microhabitat. Thus, increased mite concentrations in Tucson could be related to other building characteristics.

VI. CONCLUSIONS AND FUTURE RESEARCH GOALS

This preliminary study reports the rejection of the hypothesis that house dust mites are not present in Tucson. We expect to continue finding mites in Tucson. We hypothesize that mites will vary in the seasonal patterns reflected in these studies from year to year. We plan to sample mites from both air

conditioned and evaporatively-cooled homes and anticipate that mite populations will be greater in evaporatively-cooled homes. Since mites were collected
in some homes in numbers exceeding 100 mites/g of dust, they may affect the
health of some subjects. Once a representative sample of subjects and homes
are analyzed, we will evaluate mite skin test reactivity with mite prevalence
in homes, and assess changes in symptoms, including peak flow, during seasons with different concentrations.

ACKNOWLEDGMENTS

Portions of this work were funded by Center for Indoor Air Research
Contract #90-0003 to Mary Kay O'Rourke and Environmental Protection
Agency Contract #CR811806 to Michael D. Lebowitz.*

We are greatly indebted to the 82 families who allowed us to sample their
homes. Special thanks to P. Boyer-Pfersdorf, D. Clark, M. Ladd, S. Rogan,
C. Thomas, J. Carpenter, D. Gray, L. McKinley and E. Sorensen, our field
staff, for sampling the homes.

REFERENCES

1. Tovey, E.R., Chapman, M.D., and Platts-Mills, T.A.E., Mite faeces are a major source of
   house dust allergens, *Nature*, 289, 592, 1981.
2. Tovey, E.R., Chapman, M.D., Wells, C.W., and Platts-Mills, T.A.E., The distribution of
   dust mite allergen in the houses of patients with asthma, *Am. Rev. Respir. Dis.*, 124, 630,
   1981.
3. Arlian, L.G., Bernstein, I.L., Geis, D.P., Vyszenski-Moher, D.L., Gallagher, J.S., and
   Martin, B., Investigations of culture medium-free house dust mites. III. Antigens and
   allergens of body and fecal extract of *Dermatophagoides farinae*, *J. Allergy Clin. Immunol.*, 79, 457, 1987.
4. Arlian, L.G., Bernstein, I.L., Geis, D.P., Vyszenski-Moher, D.L., and Gallagher, J.S.,
   Antigenicity and allergenicity of body and fecal extracts of the mite *Dermatophagoides
   pteronyssinus* (Acari: Pyroglyphidae), *J. Med. Entomol.*, 24, 252, 1987.
5. Arlian, L.G., Biology and ecology of house dust mites, *Dermatophagoides* spp. and
   Euroglyphus spp., *Immunology and Allergy Clinics of North America*, 9, 339, 1989.
6. Arlian, L.G., Rapp, C.M., and Ahmed, S.G., Development of the house dust mite *Dermatophagoides pteronyssinus*, *J. Med. Entomol*, 27, 1035, 1990.
7. Arlian, L.G., Humidity requirements and water balance of house dust mites, *J. Exp. Appl.
   Acarol.,* in press.

* Disclaimer: Although the research described in this article has been supported by the United
States Environmental Protection Agency, it has not been subjected to Agency review and therefore
does not necessarily reflect the views of the Agency and no official endorsement should be inferred.
Mention of trade names or commercial products does not constitute endorsement or recommen-
dation for use.

8. Arlian, L.G., Bernstein, I.L., and Gallagher, J.S., The prevalence of house dust mites, *Dermatophagoides* spp. and associated environmental conditions in homes in Ohio, *J. Allergy Clin. Immunol.*, 69, 527, 1982.

9. Platts-Mills, T.A.E. and Chapman, M.D., Dust mites: immunology, allergic disease, and environmental control, *J. Allergy Clin. Immunol.*, 80, 755, 1987.

10. Arlian, L.G., Water exchange and effect of water vapour activity on metabolic rate in the house dust mite, *Dermatophagoides, J. Insect Physiol.*, 21, 1439, 1975.

11. Arlian, L.G. and Veselica, M.M., Effects of temperature on the equilibrium body water mass in the mite *Dermatophagoides farinae*, *Physiol. Zool.*, 54, 393, 1981.

12. Arlian, L.G., Dehydration and survival of the European house dust mite *Dermatophagoides pteronyssinus*, *J. Med. Entomol.*, 12, 437, 1975.

13. Brandt, R.L. and Arlian, L.G., Mortality of house dust mites, *Dermatophagoides farinae* and *D. pteronyssinus*, exposed to dehydrating conditions or selected pesticides, *J. Med. Entomol.*, 13, 327, 1976.

14. Moyer, D.B., Nelson, H.S., and Arlian, L.G., House dust mites in Colorado, *Ann. Allergy*, 55, 680, 1985.

15. Brown, W.G., Halonen, M.J., Kaltenborn, W.T., and Barbee, R.A., The relationship of respiratory allergy, skin test reactivity, and serum IgE in a community population sample, *J. Allergy Clin. Immunol.*, 63, 328, 1979.

16. Barbee, R.A., Brown, W.G., Kaltenborn, W., and Halonen, M., Allergen skin-test reactivity in a community population sample: correlation with age, histamine skin reactions, and total serum immunoglobulin E, *J. Allergy Clin. Immunol.*, 68, 15, 1981.

17. Quackenboss, J.J. and Lebowitz, M.D., Epidemiological study of respiratory responses to indoor/outdoor air quality, *Environ. Int.* 15, 493, 1989.

18. Lebowitz, M.D., Quackenboss, J.J. Kollander, M., Soczek, M.L., and Colome, S., The new standard environmental inventory questionnaire for estimation of indoor concentrations, *JAPCA*, 39, 1411, 1989.

19. Lebowitz, M.D. and Burrows, B., Comparison of questionnaires: the BMRC and NHLI respiratory questionnaires and a new self completion questionnaire, *Am. Rev. Respir. Dis.*, 113, 627, 1976.

20. Pepys, J., Skin tests in diagnosis, in *Clinical Aspects of Immunology*, Gell, P.G.H., and Combs, R.A.A., Eds., Blackwell Scientific, Oxford, 1968, 189.

21. WHO (World Health Organization), Dust mite allergens and asthma-A worldwide problem, *WHO Bull.*, 66, (6), 1988.

22. Platts-Mills, T.A.E., de Werk, A.L., et al., Dust mite allergens and asthma-A worldwide problem, *J. Allergy Clin. Immunol.*, 83, 416, 1989.

23. Wharton, G.W., House dust mites, *J. Med. Entomol.*, 12, 557, 1976.

24. Arlian, L.G., Bernstein, D., Bernstein, I.L., Friedman, S., Grant, A., Lieberman, P., Lopez, M., Metzger, J., Platts-Mills, T.A.E., Schatz, M., Spector, S., Wasserman, S.I., and Zeiger, R.S., Prevalence of dust mites in the homes of people with asthma living in eight different geographic areas of the U.S., *J. Allergy Clin. Immunol.*, 90, 292, 1992.

25. Arlian, L.G., Woodford, P.J., and Bernstein, I.L., Seasonal population structure of house dust mites, *Dermatophagoides* spp. (Acari: Pyroglyphidae), *J. Med. Entomol.*, 20, 99, 1983.

26. Korsgaard, J., Mite asthma and residency: a case control study on the impact of exposure to house dust mites in dwellings, *Am. Rev. Respir. Dis.*, 128, 231, 1983.

27. Platts-Mills, T.A.E., Chapman, M.D., Pollart, S.M., Heyman, P.W., and Luczynska, C.M., Establishing health standards for indoor foreign proteins related to asthma: dust mite, cat and cockroach, *Toxicol. Ind. Health*, 6, 197, 1991.

28. O'Rourke, M.K., Fiorentino, L., Clark, D., Ladd, M., Rogan, S., Carpenter, J., Gray, D., McKinley, L., and Sorensen, E., Building materials and importance of house dust mite exposure in the Sonoran Desert, in *Indoor Air '93: Proceedings of the 6th International Conference on Indoor Air Quality and Climate*, Kalliokoski, P., Jantunen, M., and Seppanen, O., Eds., Indoor Air '93, Helsinki University of Technology, 4, 155, 1993.

29. Sellers, W.D. and Hill, P.B., *Arizona Climate 1931–1972*, University of Arizona Press, Tucson, 1974.

30. Voorhorst, R., Spieksma, F.Th. M., and Varekamp, H., *House-Dust Atopy and the House-Dust Mite Dermatophagoides pteronyssinus (Trouessart 1897)*, Stafleu's Scientific Publishing Company, Leiden, Netherlands, 1968.

31. O'Rourke, M. K., Relationships between airborne pollen concentrations and weather parameters in arid environments, in *Proceedings: Aerobiology Health, Environment: A Symposium*, Comtois, P., Ed., University of Montreal, Montreal, 1989, 55.

Chapter 7

AEROMICROBIAL ANALYSES IN A WASTEWATER TREATMENT PLANT

Jacques Lavoie
Sophie Pineau
Genevieve Marchand

CONTENTS

1-56670-206-2/96/$0.00+$.50
© 1996 by CRC Press, Inc.

I. INTRODUCTION

Reported health hazards for workers employed in wastewater treatment plants include exposure to bacteria, viruses, fungi, protozoa, and industrial waste.[1-3] Endotoxins released from the cell walls of Gram-negative bacteria may produce a variety of symptoms, including fever, respiratory problems, gastrointestinal disorders, and diarrhea.[2-4] A suggested exposure limit for airborne Gram-negative bacteria is 1000 colonies/m^3.[2,5,6] In a recent report, a concentration of 7200 colonies/m^3 of air (cfu/m^3) of Gram-negative bacteria has been reported to be equivalent to the proposed occupational exposure limit for endotoxin of 30 ng/m^3, when bacteria are sampled with the Andersen impactor (Graseby Andersen, Atlanta, GA) and grown on eosin methylene blue (EMB) agar in wastewater treatment plants.[7] The fungus *Aspergillus fumigatus* has been identified in nose and throat cultures taken from individuals exposed to compost.[2,6] *A. fumigatus* is considered potentially pathogenic and may directly infect the respiratory system of immunocompromised or otherwise susceptible individuals, as well as being implicated in hypersensitivity diseases (e.g., hypersensitivity pneumonitis).[8,9]

The objectives of this study were to compare the microbial content (bacteria and fungi) in indoor and outdoor air of a wastewater treatment plant, detect the presence of abnormally high concentrations, identify potentially toxigenic species of fungi, and propose effective corrective measures. This report summarizes the aeromicrobial analyses carried out on samples taken at an urban community's wastewater treatment plant on August 9, 1990.

II. SITE DESCRIPTION

The wastewater of all municipalities of this urban community is directed to this treatment plant. Wastewater is directed from the pumping station to the pretreatment building, where self-cleaning screens remove the largest pieces of solid waste. At the screen outlets, the water is distributed to an aerated degritter, which removes sand and other heavy particles likely to sediment in the reservoir. Ferric chloride and polymers are then added, which remove phosphate and suspended particles and reduce biochemical oxygen demand. Removal of suspended particles also increases treatment efficiency. Further sedimentation of suspended matter is promoted through the slow recirculation of the wastewater in clarifying reservoirs. The sediment is then collected with bottom-scrapers and pumped to a sludge treatment center, where it is further concentrated and incinerated. In the final stage of the treatment process, the wastewater stream is chlorinated before its release into the St. Lawrence River. Overall, this treatment process decreases wastewater suspended matter by 80%, biochemical oxygen demand by 60%, the coliform bacteria count by 95%, and phosphate content by 80%.

III. METHODOLOGY

Air samples were taken both inside and outside the plant. Sampling was performed during August, when wastewater temperature is maximal (75°F). Two Andersen N–6 samplers (Graseby Andersen, Atlanta, GA) each with a different culture medium, were used to collect air samples simultaneously. Sabouraud dextrose agar (SDA, Quelab Laboratories, Montréal, Québec, Canada) was used for general counting and identification of fungi. Tryptic soya agar (TSA, Quelab Laboratories, Montréal, Québec, Canada) was used for culture of bacteria. It should be noted that these methods only reveal culturable propagules, i.e., cells, spores, and viable mycelial fragments. The sampling rate of the Andersen samplers was 28 l/min, calibrated with a Kurz hot-wire flowmeter (Kurz Instruments Inc., Carmel Valley, CA), and a sampling period of 2 min was used. The equipment was disinfected by wiping it with a 70% ethanol solution between samples. Sixty-eight air samples (34 each for bacterial and fungal analyses) were taken at 17 different sites (2 simultaneous samples per site), 16 indoor and 1 outdoor.

IV. RESULTS

Table 1 presents bacterial and fungal concentrations measured in the air (indoor and outdoor). The most prevalent bacteria identified were species of *Staphylococcus, Bacillus, Micrococcus, Pseudomonas,* and *Enterobacteriaceae.* The Gram-negative isolates are of particular interest and include *Pseudomonas* and *Enterobacteriaceae* spp.

Total concentrations and the concentration of Gram-negative bacteria could not be determined for 6 of the 17 sites, due to sample overloading. As Table 1 indicates, Gram-negative bacteria accounted for one quarter of the total colonies, varying from 9 colonies/m^3 of air (average of simultaneous duplicates) in the truck loading area to 309 colonies/m^3 at the cyclone.

Fungal recoveries for each site are also presented in Table 1. The measured concentrations exceeded, by about twofold or more, the outdoor air concentration (662 colonies/m^3 of air) at the following sites: interceptor, plenum, conveyor, sludge cakes, truck loading area, screen #8, and above and below the filter cloths.

A. fumigatus was not recovered from any of the sites. The following species were recovered and identified and have been reported to possess toxigenic as well as allergenic potentials:[11,12] *Fusarium* sp., *Penicillium verrucosum, P. brevicompactum, P. chrysogenum, P. corylophilum, P. glabrum, P. thomii,* and *P. cyclopium.* Of these, *P. chrysogenum* and *P. cyclopium* were present at relatively high concentrations, the former at the plenum (318 colonies/m^3), the latter at the clarifier (139 colonies/m^3 of air) and at screen #8 (>6000 colonies/m^3).

TABLE 1   Airborne Bacterial and Fungal Concentrations* at Various Workstations

Workstation	Bacteria (cfu/m³)		Fungi (cfu/m³)
	Gram-negative	Total	
Outdoors	64	209	662
Interceptor	**	***	1799
Plenum	**	***	3209
Conveyor	**	***	1590
Clarifier	**	***	638
Cyclone	309	1091	181
Rotary press	82	219	571
Homogenization reservoir	82	427	400
Sludge cakes	100	662	1,317
Truck loading area	9	136	1,295
Grit chamber	91	326	982
Screen #8	254	999	>6000
Screen gate	18	146	722
Upper ventilation filter	36	254	472
Lower ventilation filter	18	118	63
Below filter cloth	**	***	>7618
Above filter cloth	**	***	>7446

*   Based on the average of two simultaneous samples.
**   Unknown.
***   Sampler overloaded, >6364 colonies/m³.

V. DISCUSSION

In theory, the microbial content of air inside mechanically ventilated buildings such as office buildings should be a function of the fungal content of outside air and bacterial shedding by human occupants.[13] Concentrations and composition of these natural aerosols may vary between seasons and as a function of other environmental and building factors.[14] Increases in the quantity of fungi in indoor air relative to outdoor air, or differences between the distribution of fungal species observed in indoor and outdoor air, and the presence of bacteria not likely to be derived from normal human sources have been considered to be indicators of a potentially problematic microbial situation.[15] The studies presented here indicate an unusual exposure situation with respect to both levels and kinds of fungi and bacteria.

Due to the number of sampling sites and financial and personnel constraints, air samples were only taken once at each location. The information obtained is therefore only applicable to the period during which sampling was performed. Furthermore, statistical analysis of the concentrations was not possible with this sampling strategy, due to an insufficient number of samples.

A previous study of wastewater treatment plants has reported concentrations of culturable bacteria sampled with the Andersen impactor of 10^2 to 10^5 colonies/m³ of air (mean, 10^4 colonies/m³).[7] As Table 1 indicates, concentrations of bacteria recovered during this study are of the same magnitude. In fact, concentrations in overloaded samples probably exceed 10^4 colonies/m^3. Wastewater treatment plants seem to be comparable to compost sites and garbage-handling plants, where concentrations of microorganisms are reported to be of the same order of magnitude.[16,17]

Although it was not possible to quantify the levels of Gram-negative bacteria in the six overloaded samples, their levels can be estimated. Assuming a minimal colony count of 6364 colonies/m^3 and a Gram-negative:Gram-positive ratio of 1:4 (the same as in the other samples), levels in the six overloaded samples are likely to have exceeded Clark and Rylander's proposed limit of 1000 colonies/m^3. A study in wastewater treatment plants suggests that airborne concentrations of 7200 colonies/m^3 of Gram-negative bacteria are equivalent to 30 ng/m^3 of air of endotoxins, the proposed 8-h time-weighted average.[7] However, those levels were estimated from sampling with the Andersen impactor, using EMB agar, a Gram-negative-selective medium, rather than the TSA agar used in this study. Selective media tend to underestimate levels of the targeted organism, but allow estimates to be made in the presence of high levels of other organisms.

Although *A. fumigatus* was not recovered, high concentrations of other potentially toxigenic fungi were measured at the plenum and clarifier, and at screen #8. At this time there are no generally accepted proposals for exposure limits to toxigenic fungi. Exposure to these taxa is not synonymous with toxin exposure; mycotoxin production varies widely based on isolate, substrate, and other factors.

VI. RECOMMENDATIONS

The following were recommended to reduce the risk of microorganism exposure:

1. Because sampling carried out immediately downstream from the ventilation system filters indicated that the air delivered to the premises is of high quality (i.e., contains no excessive levels of fungi or bacteria), the air supplied by these systems can be used to dilute high concentrations of microorganisms.
2. The general maintenance of the premises should not be neglected. Regular cleaning of accumulated dust in the ventilation system can prevent the development of centers of microbial and fungal proliferation, especially the latter.[15]
3. Particular attention should be paid to the use of personal protective equipment and protective hygiene.

McCunney[1] recommends that personnel of wastewater treatment plants observe the following practices:

- Avoid putting fingers in eyes, mouth, or ears
- Wear rubber gloves when cleaning pumps, handling wastewater, etc.
- Wash hands thoroughly before eating or smoking after work
- Keep nails short
- Keep clean and dirty clothes separate
- Report and promptly treat cuts and other injuries
- Take a shower at the end of each work day

The techniques used in this study are incapable of identifying a number of viruses, protozoa, and bacteria such as *Legionella* sp. and give only a general overview of the microbial content of the sites sampled. Further in-depth studies, relying on a greater number of samples and other culture media, are indicated. It is therefore recommended that the personal hygiene measures outlined above be followed, and that either sufficient fresh air be supplied or effective breathing masks and protective goggles be worn at the plenum, the interceptor, the clarifier, the conveyor, at screen #8, and in the vicinity of the filter cloths.

REFERENCES

1. McCunney, R.J., Health effects of work at waste water treatment plants: a review of the literature with guidelines for medical surveillance, *Am. J. Ind. Med.*, 9, 271, 1986.
2. Clark, C.S., Health effects associated with wastewater treatment and disposal, *J. WPCF*, 56, 6, 625, 1984.
3. Clark, C.S., Potential and actual biological related health risks of wastewater industry employment, *J. WPCF*, 59, 12, 999, 1987.
4. Rylander, R., Snella, M.C., Endotoxins and the lung: cellular reactions and risk for disease, *Prog. Allergy*, 33, 332, 1983.
5. Seyfried, P.L., Microorganismes, parasites et endotoxines en suspension dans la section de déshydratation d'une usine de traitement des eaux usées, Recherche appliquée, *Sciences et techniques de l'eau*, 23, 3, 275, 1990.
6. Lundholm, M., Rylander, R., Occupational symptoms among compost workers, *J. Occup. Med.*, 22, 256, 1980.
7. Laitinen, S., Nevalainen, A., Kotimaa, M., Liesivuori, J., Martikainen, P.J., Relationship between bacterial counts and endotoxin concentrations in the air of wastewater treatment plants, *Appl. Environ. Microbiol.*, 58, 11, 3774, 1992.
8. Lavoie, J., *Sampling for Microorganisms in Occupational Settings*, Institut de recherche en santé et sécurité du travail, Montreal, Canada, 1990.
9. Benenson, A.S., *Control of Communicable Diseases in Man*, American Public Health Association, Washington, D.C., 485, 1985.
10. Lenette, E.H., Balows, A., Hausler, W.J., and Shadomy, H.J., *Manual of Clinical Microbiology*, 4th ed., American Society for Microbiology, Washington D.C., 1985.
11. Health and Welfare Canada, Significance of fungi in indoor air: report of a working group, *Can. J. Public Health*, 78, S1, 1987.
12. Botton, B., Breton, A., Fevre, M., Guy, Ph., Larpent, J.P., and Veau, P., *Biotechnologies, moisissures utiles et nuisibles. Importance industrielle*, Masson, Paris, 1985.
13. Burge, H.A., Approaches to the control of indoor microbial contamination, Proc. of the ASHRAE Conference, Arlington, Virginia, 1987, ASHRAE, Atlanta, GA.

14. Pineau, S. Comtois, P., The Aeromycoflora of Montreal, in *Aerobiologica, Health and Environment, a Symposium*, Comtois, P., ed., University of Montreal, Montreal, 1989.

15. ACGIH, *Guidelines for the Assesment of Bioaerosols in the Indoor Environment,* American Conference of Governmental Industrial Hygienists, Cincinnati, OH, 1989.

16. Sigsgaard, T., Bach, B., and Malmros, P., Respiratory impairment among workers in a garbage-handling plant, *Am. J. Ind. Med.*, 17, 92, 1990.

17. Lundholm, M. and Rylander, R., Occupational symptoms among compost workers, *J. Occup. Med.*, 22, 256, 1980.

Chapter **8**

AIRBORNE MICROORGANISMS IN A DOMESTIC WASTE TRANSFER STATION

Irma Rosas
Carmen Calderón
Eva Salinas
John Lacey

CONTENTS

89

I. ABSTRACT

Culturable airborne bacteria and fungi were sampled in a domestic waste transfer station in Mexico City. Close to where the waste was handled, the geometric mean concentration of bacteria was >6700 cfu/m^3 of air, of Gram-negative bacteria >460/m^3, and of fungi >4900 cfu/m^3, of which 75% were *Penicillium* spp. Concentrations of microorganisms downwind of the waste site were greater than upwind. *Salmonella* was recovered on Trypticase Soy Agar from 14% of samples. The large concentrations of Gram-negative bacteria and fungi in the waste transfer station could lead to different types of pulmonary reactions, and thus constitute a respiratory hazard to workers, and possibly also to the neighboring population.

II. INTRODUCTION

Mexico City is one of the largest cities in the world, and a major burden of the city administration is the collection and processing of about 50,000 tons of domestic waste daily. Household waste is taken to transfer stations, where it is loaded into bulk containers for transportation to landfill sites. A large proportion of the waste is putrescible and may also contain fecal and other microorganisms from human and animal sources, e.g., from disposable diapers, animal feces.[1] It is, therefore, readily colonized by bacteria and fungi. Handling such waste may result in the dispersal of fungal and actinomycete spores, bacteria, mycotoxins, and endotoxins (lipopolysaccharide from Gram-negative bacterial cell walls) into the air,[2] presenting the risk of inhalation and possible disease both in workers and in the neighboring population. Inhalation of dusts containing microorganisms and their products can result in a range of respiratory symptoms.[3-5] Sometimes, the association between symptoms and a specific agent is clear[6,7] but often many potential causal agents are present. In work environments where organic matter is handled, the risk of allergy (including both asthma and allergic alveolitis) to microorganisms is often greater than that of infection. Allergic alveolitis, particularly, may be associated with exposure to large concentrations of microorganisms (>10^6/m^3 of air) made airborne by work-related activities. Such concentrations usually greatly exceed those normally found in outdoor air.[8]

Economic, political, and environmental constraints often require the siting of municipal solid waste processing plants and sewage treatment plants within urban and suburban areas. However, the proximity of such plants to residential areas may contribute considerably to the airborne microorganism concentrations in these areas.[9,10] Exposure to large concentrations of dust during the handling of domestic waste has previously been reported in other countries.[11-13] Air in waste transfer stations in the United Kingdom has been shown to contain 10^3–10^5 cfu bacteria/m^3 of air, including 10^2–10^3 cfu Gram-negative bacteria/m³.[11,12] However, exposure of employees and local populations to emissions from waste disposal facilities has never been evaluated in Mexico. There are 15 domestic waste transfer stations operating in Mexico City and another 10 are planned. As with any industrial plant, the potential occupational and environmental health risks associated with the operation of waste treatment plants must be evaluated if methods are to be developed which will minimize related environmental and occupational health problems.

This investigation was designed to measure concentrations of both mesophilic bacteria and fungi in the air during the handling of domestic waste, and to compare these with levels outside the transfer station.

III. MATERIALS AND METHODS

A. WASTE TRANSFER STATION

The site used in this study was a roofed waste transfer station without walls, "Central de Abastos," situated in the northeastern part of Mexico City (Figure 1). The site covers an area of about 900 m², where about 2000 tons of waste per day are handled. The area surrounding the station is almost flat, with prevailing wind from the northeast. There is open pasture with small trees to the south and the west, and residential areas to the north and the east. Waste is collected from domestic and commercial premises and taken to the transfer station in specially designed vehicles. At the station the waste is compacted into bulk containers for transport to landfill sites.

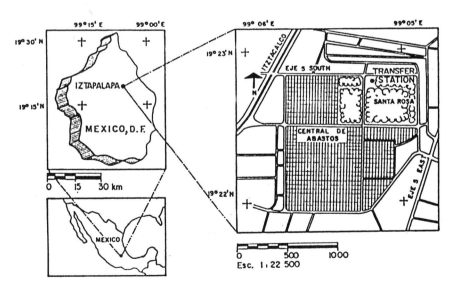

FIGURE 1   Location of sampling site.

B. SAMPLING SITES

Working environment. Sampling sites were selected to represent different work activities. These included:

Inside the station

1. Next to garbage trucks during unloading
2. Next to bulk containers during filling
3. Lunch room

Outside the station

4. 50 m downwind of the station
5. 20 m upwind of the station (Figure 1)

C. AIR SAMPLING

Air was sampled on 12 occasions at each site using two-stage Andersen samplers (Andersen Samplers, Inc., Atlanta, GA). Orifice diameters for stages 1 and 2 were 1.5 and 0.4 mm. The sampler was operated for 5 min at 28.3 l/min, and was mounted on a 2-m high tower, facing into the wind. The Andersen samplers were loaded with plastic Petri dishes containing 20 ml of either trypticase soy agar (TSA) (Difco Laboratories, Inc., Detroit, MI) for bacteria or malt extract agar (MEA) (Difco) for fungi. Meteorological parameters and the atmospheric stability were recorded outside of the station.

TSA plates were incubated at 35°C for 48 h and MEA plates at 25°C for 72 h. Colonies on each plate were counted, and counts were transformed to account for multiple deposition of particles at single impaction sites and expressed as cfu/m^3 of air.[14]

Representative fungal colonies were transferred to Difco potato dextrose agar (*Alternaria, Cladosporium*), Difco Czapek agar (*Penicillium, Aspergillus*), and Difco yeast extract glucose agar (yeast) media[15–17] and incubated at 25°C for 72 h before identification. Each sporulating fungus was identified at least to the genus level. All bacterial colonies were transferred to Oxoid violet red bile glucose agar (VRBG), and incubated at 37°C for 48 h. The suspect *Salmonella* colonies were plated on the selective isolation medium, *Salmonella Shigella* agar (SS agar Difco). Representative colonies from both VRBG and SS agar plates were isolated and identified using standard biochemical tests (ID-GNI Biotest, Biotest Diagnostics, Frankfort, Germany).

IV. RESULTS

At the transfer station, where waste was being handled, all deposition sites on the Andersen sampler plates were frequently occupied, when isolating bacteria and fungi (Table 1). Where plates were overloaded, colony counts may have been understimates so that geometric mean values are shown as being greater than (>) the indicated value. Geometric mean concentrations of particles carrying one or more viable bacteria at the three sites ranged from 5000–>9100 cfu/m^3 of air and of those carrying Gram-negative bacteria from 170–460 cfu/m^3. Geometric mean concentrations of particles carrying viable fungi ranged from 3700–>9050 cfu/m^3. Of these, 2300–7000 cfu/m^3 were *Penicillium* spp. In general, the temperature during sampling was close to the maximum temperature for the day, with light winds and unstable or neutral conditions (Table 2). Upwind and downwind concentrations were smaller than at the transfer station (Table 1). Upwind, the geometric mean concentrations of bacteria and fungi were 270 and 75 cfu/m^3, respectively. Bacteria and fungi downwind of the transfer station were more numerous (1115 and >1300 cfu/m^3, respectively) than those upwind. Gram-negative bacteria formed a large percentage (14–21%) of the total bacteria at the transfer station sites but only 1–3% outside the station. This was also the case for *Penicillium* which accounts for more than 70% of the total fungal colonies at the transfer station, but for less than 45% outside the station. Usually, more than 50% of *Penicillium* cfu were in the respirable fraction (lower stage of the Andersen sampler) (Figure 2) at both transfer station and downwind sampling sites, while most Gram-negative bacteria were in the nonrespirable fraction (top stage).

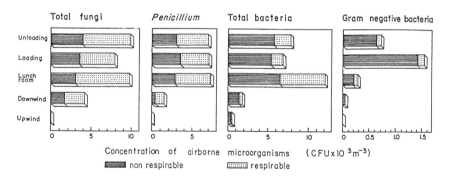

FIGURE 2   Concentration (arithmetic mean) of respirable and nonrespirable fungi *Penicillium*, total Bacteria, and Gram-negative bacteria.

The different genera of fungi that were isolated are listed in Table 3. *Penicillium* was usually the predominant colony type. However *Alternaria* and *Cladosporium* were also frequently isolated at the upwind site. Also listed in Table 3 are genera of Gram-negative bacteria that were isolated. *Enterobacter* predominated, being present in 78% of samples and forming 18% of the total Gram-negative bacteria counted. *Salmonella* was isolated from 14% of air samples in the transfer station but was found in small numbers only. Suggested limit values for concentrations of bacteria[18,19] of 2500 and 5000 cfu/m^3 of air

TABLE 1 Airborne Concentrations of Microorganisms (cfu/m³), in and around a Transfer Station

Sampling Station	Number of Samples	Total Bacteria		Gram-negative Bacteria		
		Range	Geometric Mean	Range	Geometric Mean	Percentage of Total
Unloading trucks	12	2,200–14,800[a]	>6,700	11–3,920	>270	14
Loading containers	12	350–14,800[a]	>5,000	11–4,480	>460	21
Lunch room	12	2,220–14,800[a]	>9,100	21–1,280	>170	3
Downwind	12	270–4,220	1,115	0–200	—	3
Upwind	12	60–740	270	0–14	—	1

Sampling Station	Mesophilic Fungi		Penicillium		
	Range	Geometric Mean	Range	Geometric Mean	Percentage of Total
Unloading trucks	170–14,800[a]	>4,900	28–14,800[a]	>2,300	70
Loading containers	340–14,800[a]	>3,700	63–14,800[a]	>2,650	75
Lunch room	6,020–14,800[a]	>9,050	95–14,800[a]	>7,000	79
Downwind	165–14,800[a]	>1,300	22–6,450	320	35
Upwind	20–175	75	4–100	15	45

[a] Maximum concentration threshold of Andersen sampler exceeded; geometric means are underestimates.

TABLE 2   Meteorological Parameters and Atmospheric Stability during the Sampling Period

Date	Air Temperature	Daytime Max. Temp. (°C)	Relative Humidity (%)	Wind Speed (m/s)	Atmospheric Stability Index[26]
22 Aug 89	25	26	51.4	0	1
28 Aug 89	22	25	60.0	0	2
06 Sep 89	22	22	55.1	0	2
19 Sep 89	21	22	71.7	4	3
29 Sep 89	20	22	52.5	3	2
11 Oct 89	22	22	43.5	4	3
19 Oct 89	15	17.5	60.7	0	4
27 Oct 89	21	26	47.8	0	2
06 Nov 89	22	27.5	48.8	2	2
14 Nov 89	23	26.8	39.6	0	2
23 Nov 89	17	20	62.4	2	4
01 Dec 89	17	21.2	55.8	2	4

were exceeded in 90 and 25% of samples, respectively (Figure 3), and limit values of Gram-negative bacteria[2] of 1000 cfu/m^3 were exceeded in 35% of samples.

V. DISCUSSION

This survey shows that a range of bacteria and fungi can be aerosolized when handling domestic waste. Many of the reported concentrations, especially in the transfer station are underestimates due to overloading of the Andersen plates.[20] Also, sampling was usually done during maximum daytime temperatures, when atmospheric conditions are most unstable, which could also affect recoveries.

To evaluate the contamination levels of these sampling sites, it is necesary to compare the results with published standards. Boutin et al.[18] suggested an upper limit of 2500 cfu airborne bacteria/m^3 of air was acceptable. However, 5000 cfu/m^3 has been recommended as a level indicating an abnormal source of bacteria in indoor environments.[19] With respect to Gram-negative bacteria, Rylander et al.,[2] studying endotoxin and related symptoms among compost workers, suggested a maximum of 1000 Gram-negative cfu/m^3 for safe working conditions. A large proportion of the concentrations measured within the transfer station exceeded these standards, indicating a high level of contamination.

Respiratory symptoms, abdominal pains, and diarrhea have often been reported by domestic waste transfer station workers.[11,21] These symptoms can be caused by Gram-negative bacteria, such as *Enterobacter cloacae, Escherichia coli,* and *Citrobacter freundii*, which have been shown to be toxic to experimental animals inhaling 2.5×10^9 bacteria/m^3, over a 40-min exposure

TABLE 3 **Isolation and Abundance of Identified Gram-Negative Bacteria and Fungi in Air Samples from the Transfer Station**

	Frequency of Isolation (%)[a]			Abundance (cfu/m^3)[b]		
Genera	Station[c]	Downwind	Upwind	Station[c]	Downwind	Upwind
Fungi						
Alternaria	5	17	100	2	4	20
Aspergillus	47	42	92	393	34	11
Cladosporium	17	75	83	36	400	14
Monilia	95	100	33	54	40	1
Penicillium	100	100	100	7760	2975	21
Rhizopus	78	67	17	40	32	1
Yeasts	22	42	8	262	89	1
Others	36	83	17	259	672	7
Bacteria						
Acinetobacter	70	42	—	124	7	—
Actinobacillus	17	25	—	42	2	—
Alcaligenes	39	25	20	33	2	0.2
Citrobacter	44	18	—	74	3	—
Enterobacter	78	55	20	163	12	0.5
Escherichia	58	42	20	91	6	0.2
Flavobacterium	31	8	20	36	1	0.5
Hafnia	44	33	40	11	3	1
Klebsiella	28	17	—	30	1	—
Proteus	8	—	—	14	—	—
Pseudomonas	22	8	—	16	1	—
Salmonella	14	—	—	7	—	—
Serratia	50	18	—	91	5	—
Yersinia	25	—	—	19	—	—
Others	61	25	10	103	6	0.5

[a] Percentage of samples containing each taxon.
[b] Concentration (arithmetic mean) of isolates classified within each taxon.
[c] Average of the 3 sampling sites located in the station.

period.[22,23] All three species were frequently isolated from the air of the transfer station. Moreover, enteropathogenic bacteria, such as *Salmonella*, were recovered from 14% of samples collected close to the lunch room. Its occurrence may have been related to the handling of diapers, which formed 20–30% of the bulk of the domestic waste.

Since particle size plays an important role in lung penetration and retention, the aerodynamic sizes of particles were deduced from the distribution of colonies recovered in the two-stage Andersen sampler. In general, 50 to 60% of fungal-containing particles, including *Penicillium*, were collected on the lower stage and were of a size that could penetrate into the lower airways.[24] This result was similar to that of another study done in houses.[25] Moreover *Penicillium*, a potential allergen, comprised 75% of the total airborne fungi collected during the handling of the waste. Because of the high concentrations found, we think that these bioaerosols probably constitute a risk to the health of the workers at the station and of the neighboring population.

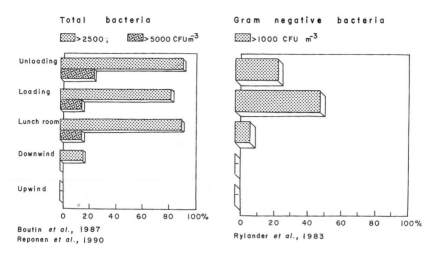

FIGURE 3 Occurrence of airborne concentration exceeding 2500 and 5000 total bacteria or 1000 Gram-negative bacteria/m³ of air.

ACKNOWLEDGMENTS

This work was partially supported by the Departamento del Distrito Federal, Dirección Técnica de Desechos Sólidos and Programa Universitario de Investigacion en Salud. We are grateful to Mr. José Juan Morales Reyes, Mrs. Alma Luz Yela Miranda, and Mr. Alfredo Rodríguez for their technical assistance.

REFERENCES

1. Peterson, M., Soiled disposable diapers: a potential source of virus, *Am. J. Public Health*, 64, 912, 1974.
2. Rylander, R., Lundholm, M., and Clark, C.S., Exposure to aerosols of micro-organisms and toxins during handling of sewage sludge, in *Biological Health Risk of Sludge Disposal to Land in Cold Climates*, Wallis, P.M. and Lohmann, D.L., Eds., University of Calgary Press, Alberta, Canada, 1983, 69.
3. Forster, H.W., Crook, B., Platts, B.W., Lacey, J., and Topping, M.D., Investigation of organic aerosols generated during sugar beet slicing, *Am. Ind. Hyg. Assoc.*, 50, 44, 1989.
4. Rosas, I., Gutierrez, S., Yela, A., Selman, M., Teran, L., and Mendoza, A., Respuesta de los trabajadores a los microorganismos suspendidos en la atmosfera de una fabrica de papel, *Arch. Invest. Med.*, 19, 23, 1988.
5. Rylander, R., Lung diseases caused by organic dust in the farm environment, *Am. J. Ind. Med.*, 10, 221, 1986.
6. Lacey, J. and Crook, B., Fungal and actinomycete spores as pollutants of the work place and occupational allergens, *Ann. Occup. Hyg.*, 32, 515, 1988.
7. Pepys, J., Jenkins, P., Festenstein, G., Gregory, P.H., Lacey, M.E., and Skinner, F., Farmer's lung, thermophilic actinomycetes as a source of "farmer's lung, hay antigen," *Lancet*, 2, 607, 1963.

8. Lacey, J., Pepys, J., and Cross, T., Actinomycete and fungus spores in air as respiratory allergens, in *Safety in Microbiology*, Shapton, D.A. and Board, R.G., Eds., Society for Applied Bacteriology Technical Series No. 6, Academic Press, London, 1972, 151.

9. Fannin, K., Vana S., and Jakubowski, W., Effect of an activated sludge wastewater treatment plant on ambient air densities of aerosol containing bacteria an viruses, *Appl. Environ. Microbiol.*, 49, 1191, 1985.

10. Millner, P., Basset, D., and Marsh, P., Dispersal of *Aspergillus fumigatus* from sewage sludge compost piles subject to mechanical agitation in open air, *Appl. Environ. Microbiol.*, 39, 1000, 1980.

11. Crook, B., Higgins, S., and Lacey, J., Airborne Gram negative bacteria associated with the handling of domestic waste, in *Advances in Aerobiology*, Boehm, G. and Leuschner, R.M., Eds., Birkhauser Verlag, Basel, 1987, 371.

12. Crook, B., Bardos, P., and Lacey, J., Domestic waste composting plants as source of airborne micro-organisms, in *Aerosols: Their Generation, Behaviour and Application*, Griffiths, W.D., Ed., Aerosol Society 2nd Conference, The Aerosol Society, London, 1988, 63.

13. Lembke, L. and Kniseley, R., Airborne microorganisms in a municipal solid waste recovery system, *Can. J. Microbiol.*, 31, 198, 1985.

14. Andersen, A.A., New sampler for the collection, sizing, and enumeration of viable airborne particles, *J. Bacteriol.*, 76, 71, 1958.

15. Dhingra, O.D. and Sinclair, J.B., *Basic Plant Pathology Methods*, CRC Press, Inc., Boca Raton, FL, 1985, 355.

16. Pitt, J., *A Laboratory Guide to Common Penicillium Species*, Commonwealth Scientific and Industrial Research Organization, Division of Food Research, North Ryde, Australia, 1985, 184.

17. Raper, K. and Fennel, D., *The Genus Aspergillus*, R.E. Klieger, New York, 1977, 686.

18. Boutin, P., Torre, M., Moline, J., and Boissinot, E., Bacterial atmospheric contamination in wastewater treatment plants, in *Advances in Aerobiology*, Boehm, G. and Leuschner, R.M. Eds., Birkhauser Verlag, Basel, 1987, 365.

19. Reponen, T., Nevalainen, A., Jantunen, M., Pellikka, M., and Kalliokoski, P., Normal range criteria for indoor air bacteria and fungal spores in a subartic climate, *Indoor Air*, 2, 26, 1992.

20. Gillespie, V.L., Clark, C.S., Bjornson, H.S., Samuels, S.J., and Holland, J.W., A comparison of two-stage and six-stage Andersen impactors for viable aerosols, *Am. Ind. Hyg. Assoc.*, 42, 858, 1981.

21. Lundholm, M. and Rylander, R., Occupational symptoms among compost workers, *J. Occup. Med.,* 22, 256, 1980.

22. Baseler, M., Fogelmark, B., and Burrel, R., Differential toxicity of inhaled Gram negative bacteria, *Infect. Immun.*, 40, 133, 1983.

23. Clark, S., Rylander, R., and Larsson, L., Levels of Gram negative bacteria, *Aspergillus fumigatus*, dust and endotoxin at compost plants, *Appl. Environ. Microbiol.,* 45, 1501, 1983.

24. Rantio-Lehtimaki, A., Evaluating the penetration of *Cladosporium* spores into the human respiratory system on the basis of aerobiological sampling results, *Allergy*, 44, 18, 1989.

25. Flannigan, B., McCabe, E.M., and McGarry, F., Allergenic and toxigenic micro-organisms in houses, *J. Appl. Bacteriol. Symp. Suppl.*, 70, 61 S, 1991.

26. Turner, D.B., A diffusion model for an urban area, *J. Appl. Met.*, 3, 83, 1964.

Chapter **9**

LEGIONELLA IN COOLING TOWERS: USE OF *LEGIONELLA*–TOTAL BACTERIA RATIOS

Richard D. Miller
Kathy A. Kenepp

CONTENTS

1-56670-206-2/96/$0.00+$.50
© 1996 by CRC Press, Inc.

I. INTRODUCTION

Legionnaires' disease (one form of legionellosis) is a type of bacterial pneumonia caused by members of the genus *Legionella*, primarily *L. pneumophila*. While other milder forms of legionellosis (i.e., Pontiac fever) may also be caused by these bacteria, Legionnaires' disease is the most frequently recognized and most serious form of the disease. *Legionella* spp. are ubiquitous in freshwater aquatic environments, although they usually occur at very low levels[1] and are frequently associated with free-living amoebae and other protozoa.[2-5] The disease is initiated following inhalation of sufficient numbers of *Legionella* in the form of an aerosol from an environmental source. While disease transmission has been documented from a variety of sources, including potable hot water plumbing systems,[6] hot tubs and whirlpools,[7] and a grocery store vegetable mister,[8] perhaps the most widely studied bioaerosol sources are water cooling towers and evaporative condensers.[9-14] As part of their normal operation, cooling towers produce a bacteria-laden aerosol mist that can be drawn into the air intakes of buildings or can shower unsuspecting passers-by.

The routine surveillance of cooling towers as an approach to the control of Legionnaires' disease remains a somewhat controversial topic. Most public health officials currently recommend against routine testing, based on the ubiquitous nature of *Legionella* in cooling towers in the absence of disease. Nevertheless, building owners, managers, tenants, and water treatment companies are faced with the ever-present threat of an outbreak of Legionnaires' disease related to their cooling towers. Sincere public health concerns, along with the legal/financial liabilities associated with an outbreak of Legionnaires' disease, warrant demands for reliable risk assessments. This issue was addressed by Morris and Feeley[15] in a presentation at the 1990 American Society of Heating, Refrigeration, and Air-Conditioning Engineers (ASHRAE) Annual Meeting, St. Louis, MO, where suggested guidelines were introduced. These recommendations, based on levels of *L. pneumophila* isolated from cooling towers associated with outbreaks of disease, propose that immediate remedial action should be taken if the level of viable (culturable) *Legionella* in a cooling tower exceeds 1000/ml. Such high levels obviously represent amplification of *Legionella* in the tower. However, lower level recoveries are more difficult to interpret.

Despite the ubiquitous nature of this bacterium in aquatic habitats, *Legionella* in environmental water samples is rarely observed in excess of 1% of the total bacterial population.[1] In contrast, preliminary results from our laboratory (unpublished data) have consistently shown that culturable *Legionella* may be observed in cooling towers in the absence of any other culturable heterotrophic bacteria, clearly indicating selection for and amplication of *Legionella*.

The studies presented here represent the analyses of nearly 800 cooling towers made by our laboratory over a 3-year period from 1988–90, and the assessment of disease outbreak potential using criteria based on both levels of culturable *Legionella* and the *Legionella*–total bacteria ratios.

II. MATERIALS AND METHODS

A. COOLING TOWER SAMPLES

All samples used in this study were obtained on a volunteer basis from cooling towers in the Baltimore-Washington, D.C. area. One-liter water samples were obtained by water treatment personnel during normal maintenance activities and were shipped unrefrigerated to our laboratory for analysis via overnight air express. Previous studies in our laboratory have indicated that levels of *Legionella* are not altered significantly during simulated conditions of such shipment, in contrast to refrigeration which often led to lowering of *Legionella* recoveries (unpublished data). Information on the specific names and exact schedule of chemical biocide additions was not obtained for individual cooling tower samples in this study. Nevertheless, it was specifically requested that samples not be taken within 24 h after biocide additions.

B. ISOLATION AND QUANTITATION OF *LEGIONELLA*

Legionella were isolated using a modification of the low-pH treatment and selective media protocols first described by Bopp et al.[16] Briefly, a 500-ml sample from each cooling tower was first filtered through a 47-mm diameter, 0.45-μm pore size Nuclepore® polycarbonate filter (Corning Costar Corp., Cambridge, MA). The filter was then placed in a small plastic jar with 5 ml of sterile distilled water and the bacteria removed by mild sonic vibration for 10 min in a bath-type sonicator. A 1-ml portion of the filter concentrated material was then acidified to pH 2.2 by addition of 1 ml of 0.2 M HCl-KCl buffer. After a 10 min period at room temperature, the sample was neutralized by addition of 1 ml of KOH. Portions (10 and 100 μl) from the acid-treated sample and the filter-concentrated material were spread-plated onto both Buffered Charcoal-Yeast Extract (BCYE) agar and Glycine-Vancomycin-Polymyxin (GVP) selective agar medium.[17] In addition, 10^{-1}, 10^{-2}, 10^{-3}, and 10^{-4} dilutions of the original sample were also plated on the same two media. All incubations were at 37°C. After 3 and 5 days of incubation, typical *Legionella* colonies were counted, and identification was confirmed by lack of growth on media without L-cysteine, as well as immunofluorescence microscopy using a monoclonal antibody reagent specific for *L. pneumophila* (Genetics Systems, Inc., Seattle, WA). The levels of culturable *Legionella* were calculated from the number of colonies on each agar plate and expressed as cfu/ml of original sample (corrected for the dilution or concentration of the sample).

C. DIRECT FLUORESCENT ANTIBODY (DFA) MICROSCOPY

A 10-µl sample of filter-concentrated cooling tower water was placed in a well (7-mm diameter) on a Teflon®-coated (E.I. duPont de Nemours & Co., Inc., Wilmington, DE) glass slide, air-dried, and heat-fixed. The samples were stained with the MERIFLUOR®-Legionella reagent from Meridian Diagnostics (Cincinnati, OH), which recognizes 33 different strains and serogroups of *Legionella*. The number of fluorescing bacteria in the specimen was determined by counting microscopically.

D. TOTAL CULTURABLE BACTERIAL COUNT

An estimate of the "total" culturable bacteria was also determined by counting all *Legionella* and non-*Legionella* colonies on the 10^{-1}, 10^{-2}, 10^{-3}, and 10^{-4} dilutions of the cooling tower samples plated on BCYE agar. The results were expressed as cfu/ml of original sample and were corrected to include any additional *Legionella* colonies detected on the GVP agar plates and acid treatment procedure. A non-*Legionella* count was also used for some comparisons and was calculated from the total number of non-*Legionella* colonies on the BCYE agar plates. The number of non-*Legionella* colonies obtained on this nonselective medium would be expected to be similar to other standard media (i.e., Trypticase soy agar, R2A agar, etc.) used for heterotrophic bacterial plate counts.

III. RESULTS

A. PREVALENCE OF *LEGIONELLA* IN THE COOLING TOWERS

As shown in Table 1, samples from a total of 794 cooling towers were examined during the 3-year period of 1988–90. *Legionella* was found in 51, 28, and 32% of towers in 1988, 1989, and 1990, respectively. All of the isolates from these positive samples were identified as *L. pneumophila*. When measured by immunofluorescence microscopy (living and dead *Legionella*), additional towers were found to be positive. Thus, there was some evidence of *Legionella* colonization in approximately 75–80% of the cooling towers.

B. QUANTITATION OF *LEGIONELLA* IN THE COOLING TOWERS

A large range of values was observed for the numbers of culturable *Legionella* in the samples (Table 2). A significant number of towers each year were in the Morris and Feeley[15] "high risk" category (greater than 1000 cfu/ml), and several were above 10,000 cfu/ml. It was observed that 29 of 107 towers (27%) in 1988 were in this high category, followed by 10 of 72 (14%) in 1989, and 19 of 103 (18%) in 1990. Most of the remainder of the towers fell into the moderate and low categories, with a smaller number that were very low.

TABLE 1   Prevalence of *L. pneumophila* in 794 Cooling Towers Sampled During a 3-Year Period from 1988–90

Group[a]	Number of Towers in Each Group (%)		
	1988	1989	1990
Culturable *Legionella*	107 (51)	72 (28)	103 (32)
DFA, positive only	51 (24)	131 (50)	157 (49)
Negative	51 (24)	59 (23)	63 (20)
Total	209	262	323

[a]   The culturable *Legionella* group contains all cooling towers that were positive for *Legionella* by both plate count and direct fluorescent antibody (DFA) immunofluorescence microscopy.

TABLE 2   Quantitation of Culturable *L. pneumophila* in 282 Cooling Towers Sampled During the 3-Year Period from 1988–90

Legionella (cfu/ml)	Risk[a]	Number of Towers in Each Group		
		1988	1989	1990
>10,000	High	3	0	1
1,000–10,000	High	26	10	18
100–999	Moderate	35	22	43
10–99	Low	29	28	34
<10	Very low	14	12	7

[a]   Based on the relative risk assessments proposed by Morris and Feeley[15] which were based on levels of *Legionella* detected in cooling towers associated with outbreaks of Legionnaires' disease.

C. QUANTITATION OF OTHER BACTERIA IN THE COOLING TOWERS

The number of culturable non-*Legionella* bacteria for each cooling tower sample was also recorded in order to determine if the numbers correlated with the levels of culturable *Legionella*. As shown in Table 3 for 1988, the range of numbers for the non-*Legionella* bacteria in the samples varied greatly from <1 cfu/ml to 5,000,000 cfu/ml. While the "average" cooling tower containing culturable *Legionella* did have fewer non-*Legionella* bacteria compared with the towers that were *Legionella*-free (or positive with DFA only), the large range of values made it impossible to predict the presence or absence (or the levels) of *Legionella* based solely on the number of other bacteria in the tower. Similar results were obtained for 1989 and 1990 (data not shown).

D. *LEGIONELLA*–TOTAL BACTERIA RATIOS IN THE COOLING TOWERS

As shown in Table 4, approximately 70% of the towers in the present study had ratios of culturable *Legionella*:total culturable bacteria that were

TABLE 3 Quantitation of Other Bacteria in 209 Cooling Towers Sampled During 1988

	Total Non-*Legionella* Count in Each Group (cfu/ml)	
Group[a]	Range	Mean
Culturable *Legionella*	<1–800,000	33,000
DFA positive only	30–5,000,000	430,000
Negative	<1–2,000,000	210,000

[a] The culturable *Legionella* group contains all cooling towers that were positive for *Legionella* by both plate count and direct fluorescent antibody (DFA) immuno-fluorescence microscopy.

greater than 1:100. In addition, *Legionella* comprised at least 50% of the total culturable bacteria in 39 of 282 tower samples over the 3-year period. Of particular concern, was the finding that 20 of the towers (Table 5) had no other culturable bacteria on the BCYE agar (i.e., 100% *Legionella*), and, of these, 13 had levels of *Legionella* in the high risk range.

TABLE 4 *Legionella* as a Percentage of Total Bacteria in 282 Cooling Towers Sampled During the 3-Year Period from 1988–90

	Number of Towers in Each Group		
Legionella (% of Total Bacteria)	1988	1989	1990
>50	14	4	21
11–50	26	13	20
1–10	38	26	30
<1	29	29	32
Total	107	72	103

TABLE 5 *Legionella* Concentrations in Individual Cooling Towers with 100% *L. pneumophila*

Total Culturable *Legionella* (cfu/ml)		
1988	1989	1990
20,000	20	10,000
10,000		2,200
8,000		2,000
2,000		1,500
2,000		1,300
2,000		1,200
400		1,000
50		500
		200
		150
		10

IV. DISCUSSION

The assessment of risk for disease outbreaks and minimization of risk based on routine surveillance of *Legionella* in contaminated cooling towers is complicated by the ubiquitous nature of this bacterium in aquatic environments, as well as the numerous factors involved in the proliferation of *Legionella* in cooling towers. These factors include the temperature and pH of the water, the availability of nutrients (organic compounds, iron, phosphate, etc.), the numbers and types of other bacteria, the presence of amoeba and other protozoa, the levels of *Legionella* in the make-up water, and certainly the nature of the biocide control program. In addition, while there is general agreement that the persistence of high levels of *Legionella* in a cooling tower may be undesirable, there is no consensus as to what constitutes a hazardous level. The action guidelines proposed by Morris and Feeley[15] have provided the first real effort to address this issue.

In the present 3-year study we found high levels (>1000 cfu/ml) of culturable *Legionella* in 20% of the 282 cooling tower samples collected from which culturable *Legionella* was recovered. In all cases, these cooling towers were recommended for immediate disinfection which effectively eliminated the *Legionella* as determined by follow-up testing (data not presented). Most of the towers with moderate levels of *Legionella* were also disinfected as a precaution.

Our results also help to dispel the general myth that a high level of *Legionella* in a cooling tower is correlated with an excessive level of other bacteria (i.e., biofouling conditions in the tower). As we noted in Table 3, predicting the status of a cooling tower based solely on the number of non-*Legionella* culturable bacterial was totally unreliable because of the wide variation in bacterial numbers in towers with or without *Legionella*. However, it should be noted that lowest mean levels of other bacteria were found in towers with culturable *Legionella*.

In fact, perhaps the most interesting finding of this study was the occurrence of high levels of *Legionella* in the near absence of any other aerobic, heterotrophic bacteria in the tower water. Fliermans et al.[1] have previously observed that *Legionella* in the natural aquatic environment rarely exceed 1% of the total bacterial population (determined by fluorescence microscopy). Thus, the high *Legionella*–total bacterial ratios noted in many of the cooling towers of this study (Table 4) may reflect relative tendencies to select for *Legionella* in the tower, perhaps as a result of variations in resistance to the chemical biocides used to treat the water. The absence of microbial competition could lead to increased *Legionella* overgrowth in the cooling tower.

Future studies that should contribute to the development of standards that could minimize the risk of exposure to *Legionella* from cooling towers include:

1. Collection of additional data from cooling towers associated with outbreaks of Legionnaires' disease.

2. Prospective studies correlating *Legionella* numbers with specific types, concentrations, and schedules of application of biocides, including the long-term effectiveness of chlorine disinfection.
3. Additional studies on the relationship between *Legionella* proliferation and the types and numbers of amoeba (and other protozoa) in the towers.
4. A more definitive classification of *Legionella* isolates in terms of overall virulence and ability to survive aerosolization.

ACKNOWLEDGMENTS

The authors gratefully acknowledge the support of RO/CO Corporation, Aqua Aire Division, Brentwood, MD, for providing financial and technical assistance. In particular, the authors recognize the significant contributions provided by Matt Mallon and Eric deLaubenfels.

REFERENCES

1. Fliermans, C. B., Cherry, W. B., Orrison, L. H., Smith, S. J., Tison, D. L., and Pope, D. H., Ecological distribution of *Legionella pneumophila*, *Appl. Environ. Microbiol.*, 41, 9, 1981.
2. Rowbotham, T. J., Preliminary report on the pathogenicity of *Legionella pneumophila* for freshwater and soil amoebae, *J. Clin. Pathol.*, 33, 1179, 1980.
3. Fields, B. S., Shotts, E. B., Feeley, J. C., Gorman, G. W., and Martin, W. T., Proliferation of *Legionella pneumophila* as an intracellular parasite of the ciliated protozoan *Tetrahymena pyriformis*, *Appl. Environ. Microbiol.*, 47, 467, 1984.
4. Newsome, A. L., Baker, R. L., Miller, R. D., and Arnold, R. R., Interactions between *Naegleria fowleri* and *Legionella pneumophila*, *Infect. Immun.*, 50, 499, 1985.
5. Wadowsky, R. M., Butler, L. J., Cook, M. K., Verma, S. M., Paul, M. A., Fields, B. S., Keleti, G., Sykora, J. L., and Yee, R. B., Growth-supporting activity for *Legionella pneumophila* in tap water cultures and implication of hartmannellid amoebae as growth factors, *Appl. Environ. Microbiol.*, 54, 2677, 1988.
6. Wadowsky, R. M., Yee, R. B., Mezmar, L., and Wing, E., Hot water systems as sources of *Legionella pneumophila* in hospital and nonhospital plumbing fixtures, *Appl. Environ. Microbiol.*, 43, 1104, 1982.
7. Spitalny, K. C., Vogt, R. L., Oriari, L. A., Witherell, L. E., Etkind, P., and Novick, L. F., Pontiac fever associated with a whirlpool spa, *Am. J. Epidemiol.*, 120, 809, 1984.
8. WHO, Legionnaires' disease: outbreak associated with a mist machine in a retail food store, *Weekly Epidemiol. Rec.*, 65, 69, 1990.
9. Glick, T. H., Gregg, M. B., Berman, B., Mallison, G., Rhodes, W. W., and Kassanoff, I. Pontiac fever: an epidemic of unknown etiology in a health department. I. Clinical and epidemiologic aspects, *Am. J. Epidemiol.*, 197, 149, 1978.
10. Dondero, T. J., Rendtorff, R. C., Mallison, G. F., Weeks, R. M., Levy, J. S., Wong, E. W., and Schaffner, W., An outbreak of Legionnaires' disease associated with a contaminated air-conditioning cooling tower, *N. Engl. J. Med.*, 302, 365, 1980.
11. Band, J. D., LaVenture, M., Davis, J. P., Mallison, G. F., Skaliy, P., Hayes, P. S., Schell, W. L., Weiss, H., Greenberg, D. J., and Fraser, D. W., Epidemic Legionnaires' disease: airborne transmission down a chimney, *J. Am. Med. Assoc.*, 245, 2404, 1981.

12. Conwill, D. E., Werner, S. B., Dritz, S. K., Bisset, M., Coffey, E., Nygaard, G., Bradford, L., Morrison, F. R., and Knight, M. W., Legionellosis, The 1980 San Francisco outbreak, *Am. Rev. Respir. Dis.*, 126, 666, 1982.
13. Klaucke, D. N., Vogt, R. L., LaRue, D., Witherell, L. E., Orciari, L. A., Spitalny, K. C., Pelletier, R., Cherry, W. B., and Novick, L. F., Legionnaires' disease: the epidemiology of two outbreaks in Burlington, Vermont, 1980, *Am. J. Epidemiol.*, 119, 382, 1984.
14. Timbury, M. C., Donaldson, J. R., McCartney, A. C., Fallon, R. J., Sleigh, J. D., Lyon, D., Orange, G. V., Baird, D. R., Winter, J., and Wilson, T. S., Outbreak of Legionnaires disease in Glasgow Royal Infirmary: microbiological aspects, *J. Hyg., Camb.*, 97, 393, 1986.
15. Morris, G. K. and Feeley, J. C., *Legionella*: impact on indoor air quality and the HVAC industry, *Abstract of ASHRAE Annual Meeting*, St. Louis, MO, 1990, 77.
16. Bopp, C. A., Sumner, J. W., Morris, G. K., and Wells, J. G., Isolation of *Legionella* spp. from environmental water samples by low-pH treatment and use of a selective medium, *J. Clin. Microbiol.*, 13, 714, 1981.
17. Wadowsky, R. M. and Yee, R. B., Glycine-containing selective medium for isolation of *Legionellaceae* from environmental specimens, *Appl. Environ. Microbiol.*, 42, 768, 1981.

Chapter **10**

CONIFER POLLEN: IS A REASSESSMENT IN ORDER?

Mary E. Pettyjohn
Estelle Levetin

CONTENTS

1-56670-206-2/96/$0.00+$.50
© 1996 by CRC Press, Inc.

I. ABSTRACT

In general, conifer pollen is not considered an important allergen source, and only certain members of the Cupressaceae, such as *Juniperus ashei* and *J. pinchotii*, are regarded as important aeroallergens. Recent clinical reports have questioned the alleged unimportance of pine pollen as an allergen source. Because of these reports and in view of the clinical importance of other conifers, we have undertaken an aerobiological and biochemical examination of this group. Airborne *Pinus* pollen was present three months of the year in Tulsa, OK. The peak concentrations for the year 1989 were found to be 42 pollen grains/m^3 at Site A (intake orifice 12 m above ground) and 557 pollen grains/m^3 at Site B (intake orifice 1.5 m above ground). For the year 1990, Site A's peak of 123 pollen grains/m^3 of air occurred on April 23 and the peak for Site B was 905 pollen grains/m^3 on May 11. Comparison of the molecular weights of pollen wall proteins of *J. ashei, J. pinchotii, J. virginiana, Cupressus macrocarpa, Pinus taeda,* and *P. echinata* showed the presence of three proteins with identical molecular weights in all species. The allergenicity and relative homology of these proteins remains to be investigated.

II. INTRODUCTION

In the order Coniferales, only the family Cupressaceae is considered to be a significant source of airborne allergens, and *J. ashei* (mountain cedar) and *J. pinchotii* (Pinchot's juniper) are considered to produce the most important aeroallergens within the family.[1] Mountain cedar pollen has been recognized since 1929 as a major winter aeroallergen source in central Texas, and is considered the most important allergen-producing species in the genus *Juniperus*.[2,3] Although there are no populations of mountain cedar in the vicinity of Tulsa, Levetin and Buck[4] reported the presence of this pollen in the Tulsa atmosphere due to transport by southerly winds from populations in the Arbuckle Mountains in southern Oklahoma and possibly Texas. Wodehouse[3] stated that the pollen of Pinchot's juniper produced allergens that were nearly as potent as mountain cedar pollen. This species is present in the grassland and the mesquite-grassland botanical zones of southwestern Oklahoma and central and western Texas.[3,5]

In 1931, Kahn[6] reported that patients with positive intradermal reactions to mountain cedar pollen extracts also tested positive with *J. virginiana* (Eastern red cedar) extracts. Five of these patients displayed satisfactory clinical relief after receiving immunotherapy with extracts of Eastern red cedar pollen. Kahn also reported that patients with positive intradermal tests for Eastern red cedar extracts also reacted to mountain cedar extracts, and these patients obtained clinical relief when treated with extracts of mountain cedar pollen. Later, Yoo et al.[7] confirmed this cross-reactivity using immunodiffusion with rabbit antisera and patient skin testing (scratch and intradermal).

Eastern red cedar is considered the most widely distributed conifer in the eastern half of the United States.[8] In Oklahoma, Eastern red cedar is abundant throughout the state except in the panhandle.[5] Despite the abundance of this species and the cross-reactivity with mountain cedar, Eastern red cedar is not considered an important aeroallergen.[9]

Yoo et al.[7] also showed that the pollen of *C. macrocarpa* (Monterey cypress) cross-reacted with those of mountain cedar pollen. According to Wodehouse,[3] *C. macrocarpa* pollen is considered to be one of the principle allergen sources in South Africa where the species was introduced. Because it is a warm climate species, occurring naturally in Monterey County, California, Monterey cypress is not sold as an ornamental in Oklahoma or other cold climate areas.[8,10]

Pinus, in the family Pinaceae, is one of the most widely distributed genera in the Northern Hemisphere.[8] Four species of *Pinus* occur in Oklahoma including large populations of *P. taeda* and *P. echinata*.[5] Farnham,[15] and Farnham and Vaida[16] recently reported that 9.7% of allergy patients had positive prick or intradermal reactions to *P. strobus* (Eastern white pine) pollen extracts. Also, Cornford et al.[17] showed evidence of cross-reactivity between *P. radiata* and *Lolium perenne* (rye grass pollen) using immunoblotting techniques. However, pine pollen is still not considered to commonly cause clinical symptoms.[3,9,11–14]

Because of the clinical importance of a few conifers, and the abundance of airborne pine pollen in some areas, we began an aerobiological and biochemical study of conifer pollen and its allergens. This preliminary report focuses on atmospheric concentrations of pine pollen in the Tulsa area and a comparison between the proteins present in pollen extracts of six coniferous species from three different genera.

III. MATERIALS AND METHODS

A. POLLEN TYPES

The pollen types chosen for this study represent those conifers with significant populations in Oklahoma: *J. ashei, J. virginiana, J. pinchotii, P. taeda,* and *P. echinata*. In addition, *C. macrocarpa* was chosen because of its reported cross-reactivity with *J. ashei*.[7] The pollen of these species (purchased from Greer Laboratories, Inc., Lenoir, NC) also represent the two distinct types of conifer pollen morphology, saccate (having "air bladders") and nonsaccate (Figure 1). Pollen of the *Juniperus* spp. and *C. macrocarpa* are nonsaccate, and range from 20–36 µm in diameter. Pollen grains of *Pinus* spp. have two sacci, the body appears spherical with a smooth exine under the light microscope, and range from 44–65 µm excluding sacci.

B. AIR SAMPLING

Air sampling was carried out with Burkard Volumetric Spore Traps (Burkard Manufacturing Co., Ltd., Rickmansworth, UK). Sampling stations

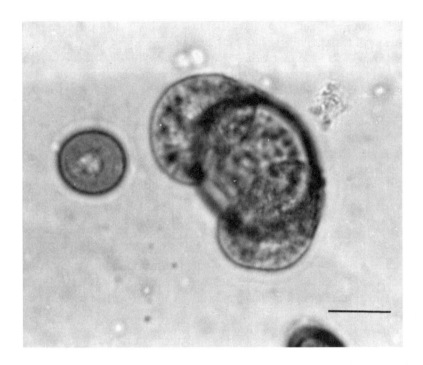

FIGURE 1 Photomicrograph of *Juniperus* and *Pinus* spp. pollen showing two types of pollen morphology present in the conifers. Bar = 20 µm.

were located on the flat-topped roof of a building at The University of Tulsa (Site A) and in a residential area 6 km southwest of the university (Site B). The intake orifice at Site A was 12 m above the ground, while that at Site B was 1.5 m above the ground. The samplers were set for 7-day sampling onto Melenex tape coated with a thin film of Lubriseal® (Thomas Scientific, Inc., Swedesboro, NJ). The tapes were changed weekly and cut into 24-h segments which were then mounted on microscope slides and stained with glycerin-jelly containing basic fuschin. One longitudinal transect was made using a 20X objective of a Nikon Lab-Photo microscope to enumerate the pine pollen. The counts were converted into 24-h average concentrations expressed as grains/m³ of air.

C. PROTEIN EXTRACTION

One gram of each pollen type was defatted with 20 ml of ether for 4 h. The ether/pollen mixtures were centrifuged and the supernatants discarded; defatted pollens were air dried for 2 h to evaporate any remaining ether. Extraction was performed by soaking the pollens in 20 ml of 0.125 M NH_4HCO_3 (pH 8.1) buffer at 4°C for 48 h with gentle shaking. The mixtures were centrifuged for 15 min, the supernatants placed in dialysis tubing with a 3500 molecular weight cut off (Spectrum Medical Industry, Inc., Los Angeles, CA), and the pollens discarded. To remove low molecular weight impurities, the extracts were dialyzed separately against fresh changes of extraction buffer for 48 h at 4°C. Following dialysis, extracts were lyophilized and stored under desiccation.

Initially the same protein extraction method was used for all pollens; however, pine pollen floated in the extraction buffer and the procedure was modified for the pine pollen by constantly stirring the pollen/buffer mixture during the 48 h of extraction.

D. PROTEIN ANALYSIS

To determine the number of proteins present and the estimated molecular weights of the proteins in each extract, sodium dodecylsulfate polyacrylamide gel electrophoresis (SDS-PAGE) was used. A 12.5% discontinuous gel was prepared based on Laemmli's[18] method and electrophoresis performed using a Bio-Rad PROTEAN® Slab Cell (Bio-Rad Laboratories, Inc. Hercules, CA). The lyophilized extracts were reconstituted in 420 µl of sample buffer (0.0625 M Tris, 10% glycerol, 2% SDS, and 5% 2-beta-mercaptoethanol), boiled at 100°C for 5 min, and analyzed for protein content using the trichloroacetic acid assay. After analysis, the samples were precipitated with equal volumes of –20°C acetone and placed in a –20°C freezer for 15 min. Next, the samples were centrifuged for 15 min at high speed in a clinical centrifuge, and the supernatants were discarded; the protein pellets were air dried to evaporate any remaining acetone. The pellets were resuspended in the appropriate amount of sample buffer to ensure loading of 100 µg of protein. The molecular weight markers were Dalton Mark VII-L® (Sigma Chemical Co., St. Louis, MO). Gels were stained with either Coomasie blue or silver stain (Sigma Chemical Co.) and molecular weights estimated using a Bio-Rad 620 Densitometer and Bio-Rad 1-D analyst software (version 2.01).

E. IMMUNOBLOTTING

After electrophoresis, the proteins from the gel were electroeluted (60 volts, 0.21 amps) onto a nitrocellulose membrane using a Bio-Rad Trans-Blot Electrophoretic Transfer Cell. Transfer was complete after 12 h. After transfer, the nitrocellulose membrane was fixed in 10% trichloroacetic acid for 1 h and the gels were stained with Coomassie blue to confirm transfer of proteins. Nonspecific binding sites on the nitrocellulose were blocked by incubating the membrane with 1% bovine serum albumin (BSA) in phosphate-buffered saline (PBS) for 1 h. After blocking, the membrane was washed three times in PBS/0.5% Tween 20 for 5 min each wash. One half of the membrane (positive control) was incubated overnight at room temperature with patient sera (diluted 1:2 in PBS/1% BSA/0.5% Tween 20) reported to contain specific IgE antibodies for mountain cedar allergens (provided by Dr. V.O. Laing, Utica Square Medical Center, Tulsa, OK). The other half of the membrane (negative control) was incubated overnight at room temperature with PBS/1% BSA/0.5% Tween

20. After incubation, the membranes were washed three times with PBS/0.5% Tween 20 for 10 min each wash. The positive and negative membranes were incubated at room temperature separately for 1 h with goat antihuman IgE conjugated with horse radish peroxidase (Sigma Chemical Co.) diluted 1:50 in PBS. After incubation, both membranes were washed three times in PBS/0.5% Tween 20 for 10 min each wash. Next the membranes were subjected to a color development process using HRP Color Reagents (Bio-Rad Laboratories, Inc.). When color development was complete, the membranes were washed three times in distilled water (5 min each) and then were allowed to air dry.

IV. RESULTS

Pinus pollen was collected on approximately 73% of the days from the first week of April through the last week of June during 1989 and 1990 (Figures 2 and 3). As shown in Figure 2, the highest concentrations for 1989 occurred during the last week of April through the first week of May. For 1990 (Figure 3), the highest concentrations were found to occur from the end of April through the middle of May. During these two years, the concentrations at Site B were higher than those at Site A. The 1989 peak concentration, 42 pollen grains/m^3, at Site A (Figure 2) occurred on April 12. Site B's peak concentration of 557 pollen grains/m^3 (Figure 2) occurred on April 28, 1989. For 1990 (Figure 3), Site A's peak of 123 pollen grains/m^3 occurred on April 23, and Site B's peak of 905 pollen grains/m^3 was on May 11.

All pollen extracts produced multiple protein bands (Figures 4 and 5). Compared to the three *Juniperus* spp. and *C. macrocarpa*, *Pinus* extracts contained a greater number of protein bands in the range of 21,000 to 97,000 Da (Figures 4 and 5). Both *Pinus* spp. produced three heavily staining bands of approximately 70,000, 48,000, and 32,000 Da. As shown in Figure 5, stirring during the extraction procedure increased the numbers of protein bands for pine pollen extracts. Extracts of the three *Juniperus* spp. and *C. macrocarpa* all produced an intensely staining band of approximately 42,000 Da (Figure 4.). Although staining intensity varied, all extracts produced three proteins with approximate molecular weights of 21,000, 32,000, and 42,000 Da.

Immunoblotting results were inconclusive. As shown in Figure 6, experimental and control blots showed identical banding patterns with the controls having a greater staining intensity. The proteins with the most intense staining were the first five molecular weight markers with marker number four (carbonic anhydrase, bovine erythrocytes) exhibiting the highest degree of staining.

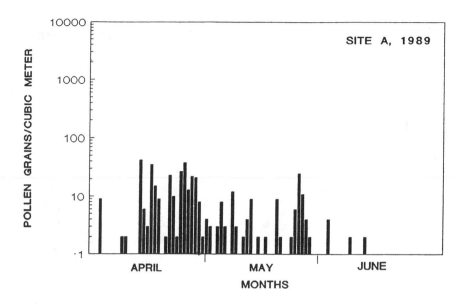

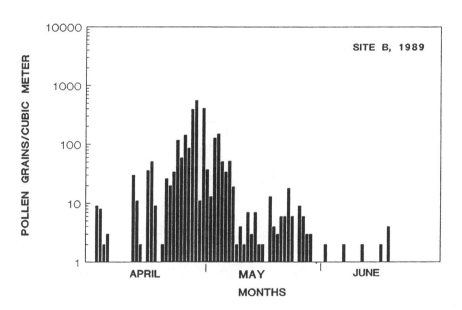

FIGURE 2 1989 Tulsa atmospheric *Pinus* pollen concentrations at Site A (intake at 12 m aboveground) and Site B (1.5 m above ground).

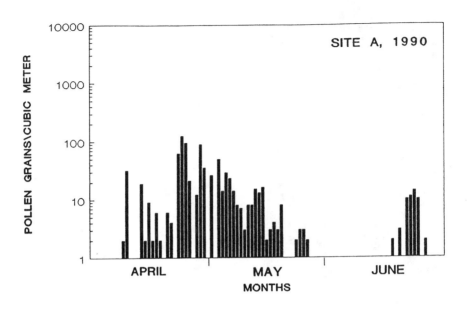

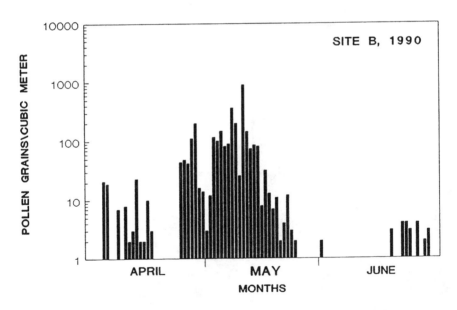

FIGURE 3   1990 Tulsa atmospheric *Pinus* pollen concentrations at Site A (intake at 12 m above ground) and Site B (1.5 m above ground). Note: data not collected from April 17 through April 24.

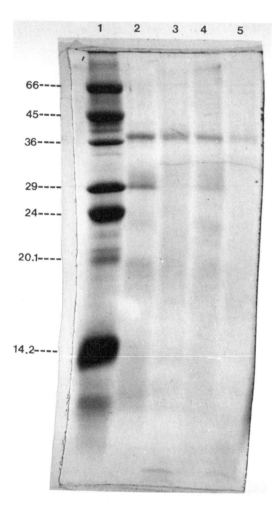

FIGURE 4 Proteins present in *Juniperus* spp. and *C. macrocarpa* following SDS-PAGE and Coomassie brilliant blue staining. Lane 1, molecular weight standards; Lane 2, *J. ashei;* Lane 3, *J. virginana;* Lane 4, *J. pinchotii;* Lane 5, *C. macrocarpa.* Molecular weights expressed in kilodaltons (kDa).

V. DISCUSSION

The genus *Pinus* is well represented in the Tulsa area. Although no major populations of native species are present within the city limits, extensive use of *Pinus* spp. in landscaping provides a source of airborne pine pollen. While saccate conifer pollen (produced exclusively by members of the Pinaceae family) cannot be easily identified to genus level with light microscopy, *Pinus* is the only member of the group native to Oklahoma. In addition, the use of

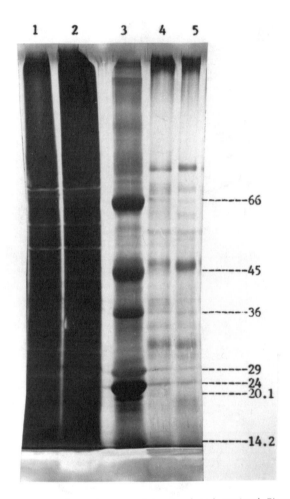

FIGURE 5 Comparison of proteins present in stirred and unstirred *Pinus* spp. pollen extracts following SDS-PAGE and silver staining. Lane 1, stirred *P. taeda*; Lane 2, stirred *P. echinata*; Lane 3, molecular weight standards; Lane 4, unstirred *P. taeda*; Lane 5, unstirred *P. echinata*. Molecular weights expressed in kDa.

other members of this family in landscaping is very limited in Tulsa. As a result, we assumed that the majority of airborne saccate pollen collected belonged to the genus *Pinus*.

Although numerous pines are present on the University of Tulsa campus, none are close to the Site A sampler which has its intake orifice at 12 m aboveground. By contrast, the Site B sampler (intake at 1.5 m) was within 15 m of a pine tree. Although Site B showed higher concentrations of pine pollen (Figures 2 and 3), all pine pollen concentrations were relatively low when compared to other tree pollen (Levetin, unpublished data). A similar study by Levetin and Buck[4] had a sampler located close to a male *J. virginiana* tree.

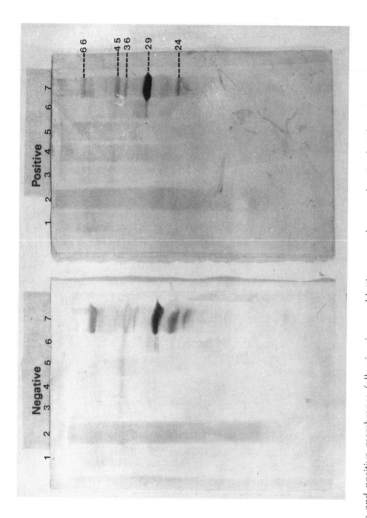

FIGURE 6 Negative and positive membranes following immunoblotting procedures and stained with HRP color development. Negative membrane incubated only with secondary antibody. Positive membrane incubated with patient sera and secondary antibody. Lanes are the same on both membranes. Lane 1, *P. echinata*; Lane 2, *P. taeda*; Lane 3, *C. macrocarpa*; Lane 4, *J. pinchotii*; Lane 5, *J. virginana*; Lane 6, *J. ashei*; Lane 7, molecular weight standards; expressed in kDa.

Their data showed the highest concentration of *J. virginiana* pollen was over 4000 grains/m³, but the highest concentration for *Pinus* pollen at Site B in the present study was only 905 grains/m³ (Figure 3). In both studies the sampler was approximately 15 meters east of the pollen source. A possible explanation for the lower pine pollen recovery in this study compared to juniper recoveries in the earlier study (*if* we assume similar source strengths) is that pine pollen settles out more quickly than juniper pollen. This is supported by the reported settling velocities of juniper and pine, 0.9 and 3.0 cm/s, respectively.[23]

Pollen extracts from the three *Juniperus* spp. and *C. macrocarpa* all produced a protein band of approximately 42,000 Da which stained intensely, suggesting a relatively large concentration. Pollen extracts from the two *Pinus* spp. had three heavily stained protein bands in common (70,000, 48,000, and 32,000 Da). The increase in numbers of protein bands produced by stirring during pine pollen extraction (Figure 5) may have been due to either degradation of large proteins or the elution of more kinds of proteins.

SDS-PAGE analysis also showed that all six species studied contained a 42,000-Da protein. According to the literature,[19–22] a 40,000 to 50,000-Da protein has been identified as the major allergen in *J. ashei* extracts. Although the attempt (via immunoblotting) to determine if the 42,000-Da protein was an allergen failed, it can be suggested that, at least for *J. ashei*, *J. virginiana*, and *C. macrocarpa*, this protein might be a common allergen and would account for the cross-reactivity found by Kahn[6] and Yoo et al.[7]

Failure of the immunoblotting procedure could have been due to several factors. Dilutions of the primary and/or secondary antibodies may not have been optimal. The identical banding patterns for both experimental and control blots probably means that the secondary antibody which was not affinity purified, may have bound to irrelevant proteins (i.e., not only to the epsilon chain of human IgE). Diluting the goat serum to lower the concentration of these nonspecific antibodies might allow resolution of those specific for IgE. Loria and Wedner[24] reported these same problems using this system for oak pollen extracts.

Recent clinical data,[15,16] the presence of airborne pine pollen, and the molecular weight similarity of pollen proteins suggest that pine pollen should not be overlooked as a possible cause of allergies. This suggestion, however, contradicts current belief that pine pollen is not allergenic.[1,3,10,13,14,17] There have been many hypotheses put forth as to why the airborne pollen of pine is rarely allergenic. Wodehouse[3] stated that humans have developed an immunity to pine pollen because pine has been around since the postglacial period. However, most allergenic plants have also been in existence this length of time. Harris and German[13] hypothesized that the waxy coating on the pollen hinders the processing of the antigen. Others suggest pine pollen is too large to enter the respiratory tract.[14] Cornford et al.[17] suggest that the low protein content of pine pollen is a reason for it being rarely allergenic. In the present study, low concentrations of airborne pine pollen were recovered within 15 m of the source, possibly indicating that *Pinus* pollen settles out of the atmosphere quickly. However, although pine pollen may not remain airborne for great distances, it still may be a potentially important allergen for residents of areas with large populations of pine trees.

J. virginiana is also not considered an important aeroallergen despite reports of cross-reactivity with *J. ashei*, its geographic abundance, and the high reported concentrations of atmospheric pollen of this species.[4,6,7,9] The biochemical results from this study and previous studies[17,20] suggest that a reassessment of the allergenicity of pollen within the order Coniferales is needed.

REFERENCES

1. Weber, R.W. and Nelson, H.S., Pollen allergens and their interrelationships, *Clin. Rev. Allergy*, 3, 291, 1985.
2. Black, J.H., Cedar hayfever, *J. Allergy*, 56, 295, 1929.
3. Wodehouse, R.P., *Hay Fever Plants*, Hafner, New York, 1971.
4. Levetin E. and Buck, P., Evidence of mountain cedar pollen in Tulsa, *Ann. Allergy*, 56, 295, 1986.
5. Levetin, E. and Buck, P., Hay fever plants in Oklahoma, *Ann. Allergy*, 45, 26, 1980.
6. Kahn, I.S. and Grothaus, E.M., Hay-fever and asthma due to red cedar, *So. Med. J.*, 24, 729, 1931.
7. Yoo, T-J., Spitz, E., and McGerity, J.L., Conifer pollen allergy: studies of immunogenicity and cross antigenicity of conifer pollens in rabbit and man, *Ann. Allergy*, 34, 87, 1975.
8. Elias, T.S., *Trees of North America,* Outdoor Life Nature Books, Van Nostrand Reinhold, New York, 1980.
9. Lewis, W.H., Vinay, P., and Zenger, V.E., *Airborne and Allergenic Pollen of North America*, The John Hopkins University Press, Baltimore, MD, 1983.
10. Strella S., personal communication, 1991.
11. Rowe, A.H., Pine pollen allergy, *J. Allergy*, 10, 377, 1939.
12. Newmark, F.M. and Itkin, I.H., Asthma due to pine pollen, *Ann. of Allergy*, 25, 251, 1967.
13. Harris, R.M. and German, D.F., The incidence of pine pollen reactivity in an allergic atopic population, *Ann. Allergy*, 55, 678, 1985.
14. Armentia, A., Quintero, A., Fernandiz-Garcia, A., Salvador, J., and Martin-Santos, J.M., Allergy to pine pollen and piñon nuts: a review of three cases. *Ann. Allergy*, 64, 49, 1990.
15. Farnham, J.E., A new look at conifer allergy, *N. Engl. Reg. Allergy Proc.,* 9, 237, 1988.
16. Farnham, J.E. and Vaida, G.A., A new look at New England tree pollen, *Proc. N. Engl. Soc. Allergy*, 3, 320, 1982.
17. Cornford, C.A., Fountain, D.W., and Burr, R.G., IgE-binding proteins from pine (*Pinus radiata* D. Don) pollen: evidence for cross-reactivity with rye grass (*Lolium perenne*), *Int. Arch. Allergy Appl. Immunol.*, 93, 41, 1990.
18. Laemmli, U.K., Cleavage of structural proteins during assembly of the head of bacteriophage T4, *Nature*, 227, 680, 1970.
19. Goetz, D.W., Vaughan, M.P., Patterson, W.R., and Reid, M.J., Monoclonal antibody identification of epitopes shared by the major allergen and other glycoproteins of mountain cedar (MC) pollen, *J. Allergy Clin. Immunol.*, 83 (Abstr.), 252, 1989.
20. Schweitz, L.A., Hylander, R.D., Whisman, B.A., Wakefield, K.J. Goetz, D.W., and Reid, M.J., Cross-reactivity among conifer pollens defined by immunoblot and monoclonal antibodies, *J. Allergy Clin. Immunol.*, 83 (Abstr.), 253, 1989.
21. Budens, R.D., Stastny, P., and Sullivan, T.J., Studies of immunochemistry and immunogenetics of mountain cedar allergy, *J. Allergy Clin. Immunol.*, 83 (Abstr.), 294, 1989.

22. Gross, G.N., Zimburean, J.M., and Capra, J.D., Isolation and partial characterization of the allergen in mountain cedar pollen, *Scand. J. Immunol.*, 8, 437, 1978.

23. Nilsson, S. and Praglowski, J., Eds., *Erdtman's Handbook of Palynology,* 2nd ed., Munksgaard, Copenhagen, 1992.

24. Loria, R.C. and Wedner, H.J., Nonspecific binding of horseradish peroxidase (HRP) labeled goat IgG to "immunoblots" of oak pollen proteins, *J. Allergy Clin. Immunol.*, 79 (Abstr.), 132, 1987.

Chapter **11**

COMPARISON OF ALLERGENIC POTENCY OF FOUR BATCHES OF *CLADOSPORIUM HERBARUM* FOR PREPARATION OF REFERENCE STANDARD

Hari M. Vijay
Maureen Burton
Gauri Muradia
N. Martin Young
Michael Corlett

CONTENTS

1-56670-206-2/96/$0.00+$.50
© 1996 by CRC Press, Inc.

123

I. ABSTRACT

Four batches (A1, A2, A3, and A4) of *Cladosporium herbarum* (IMI 96220, International Mycological Institute, Surrey, England) were grown separately on synthetic revised tobacco medium for 28 days. Extracts of the mycelia were prepared and their biochemical and immunological properties were examined. Extracts A2, A3, and A4 had 12 to 17 bands in isoelectric focusing with isoelectric point (pI) values between 3.6 to 5.9, whereas extract A1 had only 8 bands in the pI range between 3.6 to 5.4. In crossed-radioimmuno electrophoresis (CRIE) using human atopic sera, the extracts showed one dominant and three minor allergens. In IgE immunoblots of SDS-PAGE gels, each extract except A1 had strongly reactive bands at 18, 35, 41, 52, and 93 kDa molecular weight along with moderate reactivity at 15, 28, 31, 62, and 82 kDa. Extract A1 had comparatively weaker radiostained bands and the 15, 62, and 82 kDa bands appeared to be missing. The results indicate that there is a variability among the batches but a reference preparation of *C. herbarum* suitable for standardization purposes can be obtained by pooling batches of IMI 96220 isolate.

II. INTRODUCTION

Earlier, it was reported[1] that laboratory cultures of IMI 49630 and 96220 were suitable sources for a reference standard of the mold *C. herbarum*, in that they contained a representative selection of the major allergenic and antigenic components. In the present study, the batch-to-batch reproducibility of the isolate IMI 96220 was investigated.

III. MATERIALS AND METHODS

A. PREPARATION OF EXTRACTS

Four batches of IMI 96220 isolate were transferred to potato dextrose agar plates of 9-cm diameter. When the *C. herbarum* colonies were 5–6 cm in diameter, 1-cm^2 blocks were cut from the colonies and transferred to flasks containing 250 ml of synthetic revised tobacco medium[2] (prepared from a mixture of inorganic salts supplemented with vitamins). The cultures were incubated at 22°C without shaking for 4 weeks with 8-h daylight exposure every day as described earlier.[3] Each batch consisted of 10 flasks grown simultaneously. In order to avoid edge effect in the incubator, positions of the flasks were randomized. The cultures were visually monitored, dark brown or black color being regarded as the significant maturing point. The mold pellicles were collected, washed with acetone, and air dried. The pellicles from the four batches, termed A1 to A4, were individually homogenized and extracted with 1:20 w/v Tris-saline buffer, pH 7.8 for 48 h at 4°C as reported previously.[1] An additional extract, pool B, was prepared by extraction of a combination of equal amounts of the four mycelia.

B. BIOCHEMICAL ASSAYS

Protein contents of the extracts were compared by a commercial dye-binding assay (Bio-Rad Laboratories, Mississauga, Ontario), using bovine serum albumin as a standard.[4]

The extracts were screened using the Api-Zym® System (Api Laboratory Products Ltd., Quebec, Canada) which consists of a plastic gallery of 20 cupules carrying substrates and buffers for the determination of 19 enzymes: alkaline and acid phosphatases, 3 lipases, 5 peptidases, a phosphodiamidase, and 8 glycosidases.[5] Briefly, 50 ml of extract (2 mg/ml) were placed in each cupule of the device and incubated for 3 h and 45 min at 37°C, after which 20 ml of Tris-aminomethane reagent and 20 ml of Fast Blue BB reagent were added to each cupule. A reading was taken 15 min later and the reactions were graded from 0 to 5 according to the intensity of the colored reaction as compared with the manufacturer's color chart.

Analytical isoelectic focusing (IEF) was carried out in a LKB Multiphore apparatus (Model 2117) (Pharmacia Biotech, Baie d'Urfé, Quebec, Canada) using polyacrylamide gel plates with a pH range of 3.5–9.5 as reported earlier.[5] The pH markers were obtained from Pharmacia Ltd. (Montreal, Quebec, Canada). The gel plates were stained with Coomassie brilliant blue R-250 (Bio-Rad Laboratories, Montreal, Quebec, Canada).

C. IMMUNOLOGICAL ASSAYS

Previously reported procedures[6] were followed for the production of anti-sera to the mycelial extracts in New Zealand white rabbits (three rabbits per antigen) and for precipitin studies. The IgE preparation was a mixture of equal volumes of sera from six patients highly allergic to *C. herbarum* as judged by a skin test (prick) reaction of 4+ and direct RAST[6] binding of 2+ or 3+.

Crossed-immunoelectrophoresis (CIE) and crossed-radio-immunoelectro-phoresis (CRIE) were performed as described by Axelson et al.[7] and Aukrust[8] respectively, with slight modifications.[9] Briefly, 10-ml samples containing 300 mg of antigen were applied to the gel and 400 ml of rabbit antiserum against an in-house reference preparation[1] (made three years earlier than the rest of the batches of this organism) of *C. herbarum* were placed in the upper antibody gel. Electrophoresis in the first dimension was carried out at 10 volts (V)/cm for 0.5 h and in the second at 2 V/cm for 16 h. The plates were stained with Coomassie brilliant blue R-250. For CRIE, unstained CIE plates were incu-bated overnight with 400 ml of the pooled human atopic sera followed by incubation for 16–18 h with anti-human IgE-^{125}I (400,000 cpm/plate). The plates were subsequently exposed to X-ray film for 21 days at room temper-ature. Controls were included in which the atopic serum was replaced by a serum pool from six nonallergic individuals to check for nonspecific absorption of the antihuman IgE[-125]I.

Sodium dodecylsulphate-polyacrylamide gel electrophoresis (SDS-PAGE) was carried out in the Protean II® Slab Cell gel system (Bio-Rad Laboratories, Mississauga, Ontario, Canada) using the Laemmli buffer system.[10] Briefly, 15-cm square gels 1.5 mm thick were prepared with 4% polyacrylamide stacking gel in 0.125 M Tris-HCl pH 6.8, 0.1% SDS, and 12% polyacrylamide separating gel in 0.375 M Tris-HCl pH 8.8, 0.1% SDS. Fourteen to twenty-milliliter samples of the lyophilized extracts (5 mg/100 ml) containing ~40 mg protein were boiled for 5 min with an equal volume of sample buffer (0.0625 M Tris-HCl pH 6.8, 2% SDS, 5% 2-mercaptoethanol, 10% glycerol, and 0.001% bromophenol blue). The following standards were used: phosphorylase b (94 kDa), bovine serum albumin (67 kDa), ovalbumin (43 kDa), carbonic anhydrase (30 kDa), soybean trypsin inhibitor (20 kDa), and lactalbumin (14.4 kDa).

Immunoblotting was carried out as described by Towbin et al.[11] Briefly, protein bands separated on SDS-PAGE gels were electrotransferred onto nitrocellulose membrane (Bio-Rad; 0.45 mm pore size) in a Bio-Rad Trans-blot cell (Bio-Rad Laboratories). The blotting buffer was 0.02 M Tris, 0.192 M glycine, 20% methanol, and 0.01% SDS at pH 8.3. Transfer was effected at 4°C at a constant current of 200 mA for 16 h. Free protein sites on the nitrocellulose membrane were saturated by incubation with 1% bovine serum albumin at 4°C overnight. Protein bands reactive with specific IgE antibodies to *C. herbarum* were detected by incubating the nitrocellulose membrane overnight at room temperature with 5 ml of the same serum pool used in CRIE; diluted 1:10 (total volume 50 ml) in phosphate buffered saline. After several washes, the membranes were incubated with antihuman IgE-[125]I (Pharmacia Ltd.) in a final volume of 35 ml (66,000 cpm/ml) on the shaker for 16 h at room temperature and then autoradiographed.[9] Controls were included in which the atopic serum was replaced by a serum pool from (six) nonallergic individuals to check for nonspecific activity of the antihuman IgE-[125]I.

IV. RESULTS

A. CULTURE AND BIOCHEMICAL ANALYSIS

The yields of dry defatted pellicles of the batches grown from 2.5 l of the medium are given in Table 1. All batches gave similar yields, i.e., 14–21 g of dry pellicles except batch A1 which gave only 8.8 g. All four pellicle preparations had similar contents of soluble material (14.4–17.9 mg/g) although batch B yielded somewhat more (22.3 mg/g). The protein contents of the extracts were within 50% of each other ranging from 3.9–5.8% dry weight (Table 1).

The extracts contained large amounts of phosphatases, galactosidases, glucosidases, and β-glucosaminidase (Table 2). The enzyme profiles of the

TABLE 1   Yields from Various Batches of the IMI 96220 Isolate of *C. herbarum*

Extract	Defatted Mold Pellicles (g)[a]	Extracted Mold Pellicles (g)	Soluble Material[b] Yield (mg/g)	Protein %
A1	8.8	5.0	14.96	4.6
A2	21.2	5.0	14.44	4.8
A3	14.0	5.0	17.88	5.8
A4	16.1	5.0	17.24	3.9
B[c]	—	8.0	22.26	3.9

[a] From 2.5 l/cultures.
[b] From the extracted mold pellicles.
[c] Prepared by extraction of a mixture of equal amounts of the pellicles pooled from the four batches.

TABLE 2   Enzymatic Activities of *C. herbarum* Extracts

No.	Enzymes	Score A1	A2	A3	A4	B
1.	Control	0	0	0	0	0
2.	Alkaline phosphatase	3	4	4	5	5
3.	Esterase (C4)	2	2	2	3	2
4.	Esterase (C8)	2	4	3	4	4
5.	Lipase (C14)	1	2	2	3	2
6.	Leucine aminopeptidase	1	2	2	2	1
7.	Valine aminopeptidase	0	1	1	1	1
8.	Cystine aminopeptidase	0	0	0	0	0
9.	Trypsin	0	0	0	0	0
10.	Chymotrypsin	0	0	0	0	0
11.	Acid phosphatase	5	5	5	5	5
12.	Phosphodiamidase	5	5	5	5	5
13.	α-Galactosidase	2	1	1	1	1
14.	β-Galactosidase	2	5	5	4	5
15.	β-Glucuronidase	1	2	1	1	1
16.	α-Glucosidase	4	5	5	4	5
17.	β-Glucosidase	5	5	5	5	5
18.	β-Glucosaminidase	4	4	5	5	5
19.	α-Mannosidase	0	0	0	0	0
20.	α-Fucosidase	0	3	5	5	5

Note: Zero corresponds to a negative reaction; 5 to a reaction of maximum intensity.

extracts were similar with the exception of α-fucosidase which was not found in the extract A1. The extracts (including the pooled batch B) had similar isoelectric focusing patterns, i.e., 12 to 17 bands with pI values between 3.5 and 5.9, with the exception of extract A1 which had only 8 bands in the pH region 3.5–5.4 (Figure 1). The batch B had an additional band at pI 6.9 which, due to its unusual appearance, we are assuming, is an artifact.

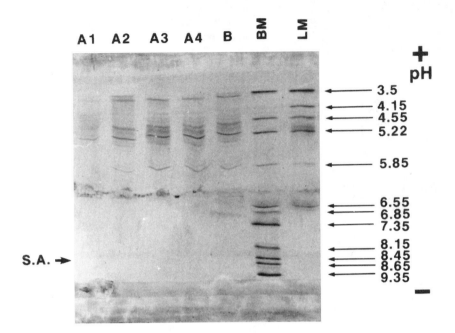

FIGURE 1   Analytical isoelectric focusing (IEF) of extracts A1, A2, A3, A4, and pool B in polyacrylamide gel containing pH 3.5–9.5 ampholines. The standard pI markers were run in the lanes marked BM (broad pI) and LM (low pI) and their values are given at the side; S.A. indicates the sample application point.

B. CROSSED-IMMUNOELECTROPHORESIS

The antigenic components of the *C. herbarum* extracts were investigated by CIE. The extracts gave complex CIE patterns with 8 to 10 precipitin arcs (Figure 2; top row) except extract A1 which had only 4 arcs. The extract B which is not shown in the Figure had also 9 to 10 bands. No cathodically migrating antigens were observed. Also, controls (normal rabbit sera) did not show any precipitin arcs.

The allergenic components among these antigens were located by incubating unstained CIE plated with human atopic serum followed by antihuman IgE^{-125}I. The autoradiographs showed (Figure 2; bottom row) that extracts A2, A3, and A4 had similar CRIE profiles, i.e., one allergen that stained intensely, and three that stained less intensely. Extract A1 had 3 allergenic components that stained with less intensity than those of the other extracts. No radiostained bands were observed when controls, i.e., nonallergic sera, were used.

C. SDS-PAGE AND IMMUNOBLOTTING ANALYSIS

The allergenic components were further characterized by SDS-PAGE followed by immunoblotting with the atopic serum pool and antihuman IgE^{-125}I (Figure 3). The autoradiographs showed that each extract had strongly reactive

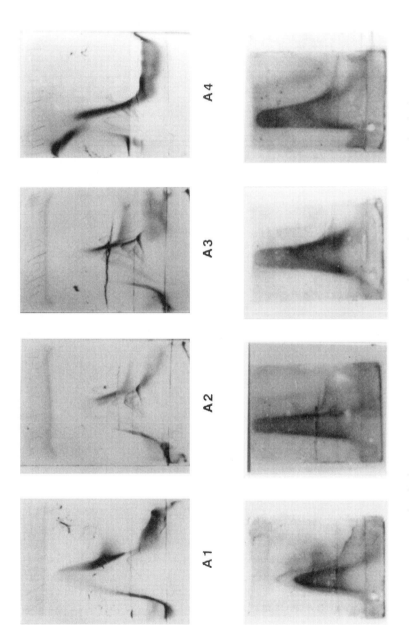

FIGURE 2 Top row: CIE patterns of *C. herbarum* extracts against antiserum to an in-house reference preparation of *C. herbarum*. Bottom row: Autoradiographs showing the CRIE patterns of duplicate unstained CIE plates.

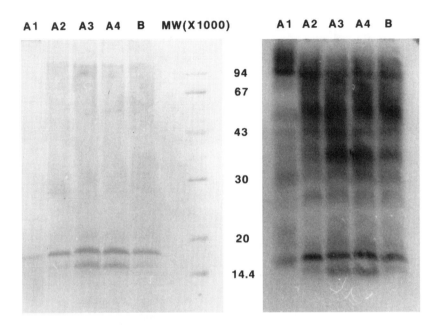

FIGURE 3   SDS-PAGE of the extracts. On the left, Coomassie brilliant blue R-250 staining of the extracts A1–A4 and B, and molecular weight standards. On the right, an autoradiograph of an immunoblot of unstained gel after incubation with the atopic serum pool followed by antihuman IgE^{-125}I.

bands at 18, 35, 41, 52, and 93 kDa molecular weight along with moderately reactive bands at 15, 28, 31, 62, and 82 kDa. Extract A1 gave comparatively weaker radiostained bands and the 15, 62, and 82 kDa bands were not seen. Controls, i.e., nonallergic sera, did not show any radiostained bands.

V. DISCUSSION

As stated above, a reference preparation of *C. herbarum* should be made from an isolate of this organism which produces all the recognized allergenic components and be of high potency and stability. In addition to these specifications, batch-to-batch reproducibility of the isolate is also of prime importance for the homogeneity of the reference standard, as has been described for an *Alternaria alternata* reference preparation,[12] so that the standard can be prepared from any single batch of the isolate. However, the findings of the present study on various batches of *C. herbarum* isolate IMI 96220 indicated some variability among the batches.

Three of the four extracts and the pooled batch B had similar IEF and enzymatic patterns, while batch A1 produced weaker staining bands in IEF and lacked fucosidase activity in the Api-Zym® test system. The band at pI

6.9 in batch B could be an artifact since this band was not observed in the original batches from which the batch B was prepared (by pooling equal amounts from four batches). It is not yet established which enzymes, if any, of *C. herbarum* are allergenic. Correlation of enzymatic activity with allergenic activity would not prove that a particular enzyme is an allergen, as the enzyme and allergen could be coexpressed, causing their levels to be linked.

In order to determine if this biochemical variability correlated with the antigenic and allergenic properties of the extracts, a battery of immunological tests, CIE, CRIE, and immunoblotting were carried out. The extracts had similar CIE patterns, i.e., 8 to 10 precipitin arcs, except batch A1 which gave only 4 arcs. In CRIE, each extract, except A1, showed one intensely staining allergen and three that stained less intensely. Intensity of staining may relate either to the concentration of a particular compound or to its relative allergenicity (i.e., number of epitopes available to bind specific antibodies). Immunoblots of the extracts A2, A3, A4, and the pool B showed strong reactivity with specific IgE antibodies to 18, 35, 41, 52, and 93 kDa molecular weight bands along with moderate reactivity to 15, 28, 31, 62, and 82 kDa bands. A1 had comparatively weaker radiostained bands and the 15, 62, and 82 kDa species were missing. The origin of this variability could be small differences in culture or extraction conditions, which, however, were maintained as uniform as feasible for all four batches. De Vries[13] reported that one cannot expect to find the same degree of germination or growth at a fixed time from batch to batch when growing *Cladosporium*.

This variability among batches of *C. herbarum* makes it necessary to prepare reference samples for standardization purposes by pooling several large batches, in the manner of extract B. Such a sample could also serve as starting material for purifying the major and minor allergens of the organism which would provide a basis for more exact characterization of crude extracts in the future.

REFERENCES

1. Vijay, H.M., Burton, M., Young, N.M., Copeland, D.F., and Corlett, M., Allergenic components of isolates of *Cladosporium herbarum, Grana*, 30, 161, 1991.
2. Shafiee, A., New synthetic medium for the production of Alternaria allergens, *Ann. Allergy,* 40, 220, 1978.
3. Vijay, H.M., Young, N.M., Jackson, G.E.D., White, G.P., and Bernstein, I.L., Studies on *Alternaria* allergens. V. Comparative biochemical and immunological studies of three isolates of *Alternaria tenuis* cultured on synthetic media, *Int. Arch. Allergy Appl. Immunol.,* 78, 37, 1986.
4. Bradford, M., A rapid and sensitive method for the quantitation of microgram quantities of protein utilizing the principle of protein-dye binding, *Anal. Biochem.*, 72, 248, 1976.
5. Vijay, H.M., Huang, H., Young, N.M., and Bernstein, I.L., Studies on *Alternaria* allergens. IV. Comparative biochemical and immunological studies of commercial *Alternaria tenuis* batches, *Int. Arch. Allergy Appl. Immunol.*, 74, 256, 1984.

6. Vijay, H.M., Huang, H., Young, N.M., and Bernstein, I.L., Studies on *Alternaria* allergens. II. Presence of two related antigens with contrasting allergenic properties in *Alternaria tenuis* extracts, *Int. Arch. Allergy Appl. Immunol.*, 65, 410, 1981.

7. Axelsen, N.H., Krøll, F., and Weeke, B., A manual in quantitative immunoelectrophoresis: methods and applications, *Scand. J. Immunol.*, 2 (Suppl. 1), 71, 1973.

8. Aukrust, L., Crossed-radioimmunoelectrophoresis studies of distinct allergens in two extracts of *Cladosporium herbarum*, *Int. Arch. Allergy Appl. Immunol.*, 58, 375, 1979.

9. Vijay, H.M., Burton, M., Young, N.M., Corlett, M., and Bernstein, I.L., Comparative studies of allergens from mycelia and culture media of four new strains of *Alternaria tenuis*, *GRANA*, 28, 53, 1989.

10. Laemmli, U.K., Cleavage of structural proteins during assembly of the head of bacteriophage T4, *Nature,* 227, 680, 1970.

11. Towbin, H., Staehelin, T., and Gordon, J., Electrophoretic transfer of protein from polycrylamide gels to nitrocellulose sheets: procedure and some applications, *Proc. Natl. Acad. Sci. U.S.A.,* 76, 4350, 1979.

12. Vijay, H.M., Burton, M., Hughes, D.W., Copeland, D.F., and Young, N.M., Comparison of allergenic potency of five batches of *Alternaria alternata* for preparation of reference standard, *Aerobiologia,* 6, 159, 1990.

13. De Vries, G.A., *Contribution to the knowledge of the genus* Cladosporium *link ex Fr. Cramer,* Baarn, Centraalbureau voor schimmetcultures, Rijks Universiteit te Utrecht, 1952.

Chapter **12**

THE DETECTION OF AIRBORNE ALLERGENS IMPLICATED IN OCCUPATIONAL ASTHMA

John Lacey
Brian Crook
A. Janaki Bai

CONTENTS

1-56670-206-2/96/$0.00+$.50
© 1996 by CRC Press, Inc.

I. ABSTRACT

The investigation of occupational asthma requires the detection of both viable and nonviable allergen-bearing particles. Isolation and enumeration of microorganisms can be achieved using filter aerosol monitors, liquid impingers, Andersen cascade impactors, and other samplers. Detection of defined allergens can be achieved only through immunoassay of samples collected, for example, by filtration, impingement, or electrostatic precipitation. Collection and extraction methods have been compared using radioallergosorbent test (RAST) inhibition and enzyme-linked immunosorbent assay (ELISA). Coca's mineral salt solution extracted fly antigen better than citrate borate buffer and phosphate buffered saline, but best recovery of egg albumin was obtained with ammonium bicarbonate. High-volume filtration samplers yielded less allergens than large-volume Litton-type electrostatic samplers which collected into liquid. Recovery on dry filters decreased with time of sampling suggesting adsorption of allergens onto the filter or its denaturation through dehydration. Yields of airborne scampi protein, measured by ELISA, were almost 20 times greater in samples collected by electrostatic precipitation than by filtration and only the liquid samples yielded allergens detectable by RAST.

II. INTRODUCTION

The aerobiological study of occupational lung disease has been primarily concerned with microbial antigens,[1] for instance, the roles of the actinomycetes *Saccharopolyspora rectivirgula* (*Micropolyspora faeni*) and *Thermoactinomyces* spp. in farmer's lung disease,[2,3] and *Thermoactinomyces sacchari* in bagassosis,[4] and of the fungus *Penicillium glabrum* (*P. frequentans*) in suberosis.[5] However, more recently, the role of allergens borne on nonviable particles including nonmicrobial allergens and antigens has become recognized. Examples include occupational asthma caused by enzymes derived from bacteria for use in biological washing powders,[6] mites in domestic and agricultural buildings,[7] fly and locust protein in insect rearing rooms,[8,9] beet protein in sugar beet factories,[10] and scampi protein in seafood factories.[11] Methods have long been established for the detection, isolation, identification, and enumeration of viable microbes that contain antigens but methods for identifying the allergens themselves in environmental samples are still evolving. Although methods for collecting both viable (culturable microorganisms) and defined antigens operate on the same principles, individual methods differ greatly in their suitability for one purpose or the other, in the assay methods they allow and in their particle collection efficiencies (Table 1).[12-14] This chapter reviews the methods used to collect and detect microorganisms and antigens on other kinds of particles and describes recent work on the development of methods for recovering and identifying allergens and antigens.

TABLE 1 Appraisal of Sampling Methods for Airborne Allergens and Antigens

Sampler	Advantages	Disadvantages
Settle plates	Inexpensive, easy	Not volumetric
Sedimentation slides		Favor large particles
Andersen culture samplers	No sample handling	Easily overloaded
	Volumetric	Use of different selective media
	Size selective	requires sequential samples
	Good for actinomycetes	
Cascade impactors	For microscopy	Easily overloaded
	Volumetric	
	Size selective	
Automatic volumetric spore trap	Continuous operation	Easily overloaded
	Volumetric	Designed for large-spored plant pathogenic fungi
	For microscopy	Trapping efficiency begins to diminish below about 10 µm
		Large mass causes sampler to lag behind changes in wind direction.
Rotating arm impactors	For microscopy	Speed of rotation must be known for volumetric use
	Volumetric	Pattern of air movement past the rods may change with configuration of the sampler
	Portable, simple to set up and use, easily constructed	Sampling efficiency for commercially available model diminishes for particles smaller than about 20 µm
	Independent of wind speed and direction	No time discrimination
	Useful for spot and intermittent samples	
Liquid impingers	Sample time unlimited	Sample handling required
	Volumetric	Jet may damage cells
	Cell counts	Difficult to collect hydrophobic particles
	Different media can be used for plating a single sample	
	No dehydration stress	
	Multi-stage version size selective	
Reuter Centrifugal	Portable, easy to operate	Sample rate uncertain
	No sample handling	Inefficient for small spores
	Particle counts	Easily overloaded
Cyclone	Sample time unlimited	Characteristics depend on design and flow rate
	Volumetric	
Electrostatic	Sample size unlimited	Sample requires handling
	High sampling rate	Corona may affect viability
	Volumetric	
Filtration	Adaptable in application (microscopy or culture), size, flow rate	Sample requires handling
	Volumetric	Dehydration may damage cells

III. METHODS FOR MICROBIAL ANTIGENS

A. SEDIMENTATION SAMPLERS

The earliest studies of allergens utilized settle plates (for culture of microorganisms) and sedimentation slides (for microscopic assessment of pollens and spores) without a consideration of how much their collection efficiencies depended on particle size and how much they were affected by wind speed and turbulence. Particles settle at a rate proportional to the square of their radius so that the occurrence of small particles, including, conidia, ascospores, and basidiospores, was greatly underestimated. At low wind speeds, shadows with little deposition occur close to the leading edge of the dish or slide. These shadows increase in size as wind speed increases until there is no deposition, unless the wind becomes turbulent, in which case as many particles can be collected on the bottom of the collection surface as on the top. Many spores and other particles can be deposited on exposed agar plates as a result of the turbulence created when the lids of Petri dishes are removed and replaced.[15]

B. IMPACTORS

Cascade impactors have been used in many occupational environments where there may be many small spores, such as actinomycetes with diameters 1 μm or less. There are many variations in design and flow rate. Models are available for both personal and area sampling for microscopic assessment. Some allow or have been modified to allow assessment by culture. All have sequential stages, each with a trapping surface behind, which causes the airstream to turn at right angles. Holes on successive stages decrease in size, accelerating the airflow so that successively smaller particles are impacted onto the trapping surface. The stage on which particles are recovered can be used to assess their aerodynamic size and how far they will penetrate into the respiratory system. Holes may be rectangular or circular, and vary in number up to 400 per stage. Collection may be onto coated glass slides (allowing microscopic assessment) or onto plastic disks or agar media (allowing culture either of washings or directly), depending on the model. Sampling rates range from 2–28 l/min.

Quantitative microscopic assessment of specific spore types is limited by the precision with which each may be identified. Morphological groups may comprise many species which have similar spores. For example, the *Aspergillus/Penicillium* group includes small spherical colorless or nonbrown spores. Bacteria and actinomycete spores cannot be distinguished microscopically without special straining. Because few organisms can be identified to species on the microscopic characteristics of the spores alone, identification requires isolation in culture, for which a range of methods are available. The Andersen sampler[16] (Graseby Andersen, Smyrna, GA) is a type of cascade impactor that deposits spores directly onto one or more Petri dishes (six in the original version) with each dish collecting particles of a different size fraction. However, in highly contaminated occupational environments, plates easily become overloaded, decreasing the accuracy of particle concentration determinations. Methods have been described for overcoming this problem by homogenizing the agar or by sampling onto a gelatin medium which is melted and then diluted before replating.[17,18] These methods allow different media to be inoculated from a single sample, but decrease the convenience of collecting microorganisms directly onto agar. The method also is likely to result in a cell or spore count rather than a particle count (spore or cell aggregates or particles, e.g., skin squames carrying cells or spores). Thus, results are not comparable with directly incubated plates. Andersen samplers have also been used for detecting airborne antigens by incorporating antisera into the agar media and observing precipitin reactions,[19] and by extracting allergens from the agar for immunoassay.

Hirst[20] based his automatic volumetric spore trap on stage two of a four-stage cascade impactor.[21] Air is sampled at 10 l/min and particles are collected, in different versions, on a glass slide or tape, coated with petroleum jelly, silicone grease or other adhesive, which moves at 2 mm/h past the intake orifice. The trapping surface in different versions is changed after 24 h or after 7 days and the catch is then classified and counted microscopically. The Hirst-type (e.g., Burkard, Burkard Manufacturing Co., Ltd., Rickmansworth, U.K.) spore traps have revolutionized our concept of the normal air spora, drawing attention particularly to the abundance of ascospores, basidiospores, and the ballistospores of *Sporobolomyces*. Use of the traps allow diurnal periodicities and changes in the spore content of the air with changing weather to be determined. Despite relatively poor collection efficiency for large spores (caused by slow response to changes in wind direction[22]) and low trapping efficiency for very small spores, Hirst-type traps do document large changes in air spora.

C. ROTATING ARM SAMPLERS

Rotorod samplers are easily portable and can be used for spot samples or intermittently during longer sampling periods. However, the airflow pattern past the trapping surface is likely to differ between models and this may affect trapping efficiency. With U-shaped rods open at the top, air is drawn into the center of the rotating rods from above and then discharged outwards past the trapping surface. However, if the rods are suspended below a bar or if there are swing shields, this pattern is likely to be modified although the nature of these changes has never been described. Air is sampled at about 110 l/min by a sampler with rods 0.16 × 6.0 cm rotating at 2300 rpm. The collection efficiency for pollens and *Puccinia graminis* urediniospores (20 μm aerodynamic diameter) is 90–95% and for *Calvatia gigantea* spores (4–5 μm) is 19%.[23] As used by most allergists, the rotorods are suspended beneath a rain shield and retract during periods of inaction.

D. CENTRIFUGAL ACCELERATION

The Reuter Centrifugal Sampler (RCS) (Biotest, Birmingham, U.K.) is convenient to use and extremely portable. However, its particle collection efficiency is difficult to determine. The instrument is calibrated to allow sampling for fixed periods from 0.5–8 min. Air is drawn into the sampler by an impeller rotating at 4096 rpm which accelerates particles in the air centrifugally so that they are deposited on an agar strip lining the circumference of the trapping chamber. The sampling rate has been reported as 210 1/m[24] and 280 l/min (manufacturer), although an "effective" sampling rate of 40 l/min (derived using a theoretical equation) is used for calculations. The trapping efficiency for particles declines as size decreases below 15 µm aerodynamic size and the low "effective" sampling rate, quoted above, is necessary to account for the low recovery of bacteria found in practice.[24] A new version of the RCS sampler has been produced to overcome the criticism that the original model cannot be calibrated but few evaluations of this new device have so far been published.

The more traditional cyclone samplers allow unlimited sampling times and, with different versions, low or high sampling rates (2–1000 l/min). Air is drawn into a conical chamber through a tangentially mounted inlet, descends in a spiral to the base of the chamber, and is then drawn out through a central outlet in the top. Particles are thrown out onto the walls of the cyclone chamber and descend into a collecting vessel at the base. The performance of cyclones depends on their geometry and sampling rate, and they can be designed to collect small particles efficiently. Small personal sampler versions, sampling at 1.9 l/min, have been used to separate respirable from nonrespirable material in the breathing zone of workers, while larger versions, sampling at about 600 l/min, have been used to isolate microorganisms and could potentially be used for collection of allergens and antigens. For bioaerosols, a liquid spray may be incorporated in the inlet of a large volume cyclone, to wash the walls and concentrate the particles contained in 500 l of air into 1 ml liquid. Catches may be assayed by culture, *Limulus* amoebocyte lysate (LAL) endotoxin assay, or by immunoassay.

E. IMPINGERS

Other methods for sampling viable microorganisms give cell rather than particle counts as a result of disruption of particles during collection or subsequent handling. In some versions of the liquid impinger, where the jet also functions as a flow limiting orifice, fragile cells may be damaged by the shear forces through the jet and by the force of impaction into the liquid. Samples are collected at 0.3–28 l/min, commonly 11–12.5 l/min, depending on the version. Multi-stage liquid impingers[25] treat cells more gently than single stage versions and give size separations that compare with those in different regions of the human respiratory system. Models are available in various sizes that sample at 10, 20, or 55 l/min, and a metal version is now available commercially (Burkard Manufacturing Company, Ltd., Rickmansworth, U.K.).

F. FILTRATION SAMPLERS

Filtration samplers also allow unlimited sampling times and a wide range of sampling rates with different versions, from 2–1000 l/min. Filtration samplers are flexible in their applications and can be used for a range of purposes. For instance, filters can be used to determine dust levels (measured gravimetrically), microorganisms (assessed by light or scanning electron microscopy or by culture), endotoxins (determined by LAL assay), antigens (assayed immunochemically), and airborne mycotoxins (assayed chemically or by immunoassay). Because the catch on filters can be resuspended and diluted before plating, samplers consisting of cassettes loaded with polycarbonate filters (Nuclepore®), powered by battery-operated personal sampler pumps, have been invaluable in highly contaminated environments for the isolation of microorganisms,[17,26,27] especially fungi, actinomycetes, and Gram-positive bacteria. However, it is possible that many Gram-negative bacteria are damaged by dehydration on the filter during sampling, resulting in underestimates of their numbers. Filtration has been used in high-volume samplers, operating at about 600 l/min, for the collection of airborne antigens. Agarwal et al.[28,29] first used filtration to collect ragweed pollen and *Alternaria* allergens which were then eluted from the glass fiber filters and detected by immunoassay.

G. ELECTROSTATIC SAMPLERS

Electrostatic samplers also allow high volume collection of air samples into liquid while recirculation of the liquid in the sampler allows concentration of airborne contaminants. Thus, problems of dehydration stress are avoided although the corona discharge might possibly affect viability of microorganisms. Litton-type electrostatic samplers (Sci-Med Inc., Eden Prairie, MN) can be operated at up to 1 m³/min although 600–700 l/min is more usual. The catch can be cultured but has also been used for immunoassay of airborne antigens.

IV. DETECTION OF NONMICROBIAL ANTIGENS

Not all airborne allergens are microbial in origin. Some may be derived from insects or mites and others may consist of aerosols of plant or animal proteins. Filtration samplers, especially high-volume samplers operating at about 600 l/min, have been used for the collection of allergens from domestic and laboratory animals, insects, plant proteins, and bacterial enzymes, which have then been detected by immunoassay.[8–11,30–32] RAST inhibition assays have generally been used for measuring reactions with specific IgE antibodies and ELISA with IgG or IgE antibodies.

The value of filtration for antigen sampling was demonstrated by Forster et al.[10] in a sugar beet factory where, in addition to bacteria, sugar beet protein was present in the air. Sera from exposed workers with respiratory symptoms reacted more often in immunodiffusion tests to the protein and to filter extracts than they did to the bacteria (Table 2). Lines of identity between beet and filter extracts indicated the presence of common antigens in both extracts. The capacity of filter extracts to inhibit ELISA tests showed that about 0.1% of the filter extracts comprised beet antigen.

TABLE 2   Antibody Responses to Airborne Contaminants in a Sugar Beet Factory[10]

| | | | | No. with Specific IgG Against | | |
| | | | | *Bacillus* spp. | *Leuconostoc mesenteroides* | *Pseudomonas* spp. |
No. of Workers	No. with Respiratory Symptoms	Beet Extract	Filter Extract			
12	9	6	5	4	2	3

Note: One worker not tested.

Methods other than filtration can also be used. In tests in a factory where scampi were deshelled using water jets, not only did a Litton-type large-volume electrostatic air sampler satisfactorily collect airborne antigen into liquid, but recovery was better than by collection on glass fiber filters at similar sampling rates (700 l/min). Five successive 60-min samples were collected simultaneously with the two types of collectors and samples were then extracted, dialyzed, freeze-dried, and then reconstituted for RAST inhibition tests. Allergen recovery from the filters was more unpredictable and much lower than from the large-volume electrostatic air sampler (often insufficent for dose-response regression analysis) (Table 3). The large-volume electrostatic air sampler revealed 3.7–8.8 μg of scampi allergen/m^3 air with smaller concentrations during the lunch hour and at the end of the shift. The poor performance of the filters could be due to failure to collect the antigen on filters (although 100 μg protein was present in the filter extracts) binding of antigen to the filter fibers, or denaturation of the antigen, perhaps as a result of dehydration.

V. COMPARISON OF METHODS FOR DETECTING AIRBORNE ANTIGEN

There have been few studies comparing the advantages of different methods of allergen collection and assay. We have recently compared methods for

TABLE 3 Comparisons, Using RAST Inhibition Assay, of Large Volume Electrostatic Air Sampler (LVAS) and Filter Sampler for Collecting Scampi Allergen[11]

LVAS			Filter		
Percent Inhibition	Sample Size (m³ of air)	Conc. Allergen (μg/m³)	Percent Inhibition	Sample Size (m³ of air)	Conc. Allergen (μg/m³)
41	42.0	3.7	6	39.0	0.4[a]
56	44.8	7.6	32	41.6	2.6
54	45.5	6.9	14	42.3	0.9[a]
44	42.0	4.4	15	39.0	1.0[a]
59	42.0	8.8	1	39.0	0.1[a]

[a] These data should be treated with caution because they fall outside the exponential plot of the calibration curve which is the best area for accurate measurement. They are therefore likely to be subject to large errors.

sampling airborne antigen in a model system using egg albumin aerosols in exposure chambers and an ELISA assay system utilizing egg albumin polyclonal antibody (Sigma Chemical Co., Poole, Dorset, UK) to further study the reasons for the differences between filter and large-volume electrostatic air samplers and to attempt to improve sensitivity of the methods. Microtiter plates were coated overnight with sample extracts or with different concentrations of egg albumin and then incubated with rabbit anti-chicken egg albumin antibody for 30 min at 30°C followed by goat anti-rabbit IgG-horse radish peroxidase conjugate for 30 min. 3,3′,5,5′-Tetramethylbenzidine was added as substrate and the enzyme reaction was allowed to proceed for 15 min, then was stopped with 10% H_2SO_4. Absorbance was measured at 450 nm. Using this procedure, 10–250 ng albumin/ml could be detected when the antibody was diluted 1:10,000 and 1–25 ng/ml when diluted 1:1000.

Ammonium bicarbonate extraction buffer allowed much better recoveries from filters than Coca's solution. Cutting and soaking filters only gave slightly better recovery than elution in chromatography tanks. Collection was found to be most efficient using the Litton-type air sampler but high recoveries were also obtained with a high-volume filtration sampler collecting onto glass fiber filters and with the Sierra-Marple eight-stage personal cascade impactor collecting onto slotted plastic discs (Table 4). Recovery of albumin atomized into the samplers was about 2.5 times greater from the electrostatic sampler than it was from the filtration sampler. Interestingly, there was little change in recovery from the filters with sampling times from 10–180 min. In further wind tunnel tests, recovery by filtration was about 70% of that by electrostatic sampler. There was little difference between different filter materials. Liquid impingers and personal monitors allowed recovery of much less albumin, especially if used closed-face. There was also little difference in recoveries between Porton raised impingers and multi-stage liquid impingers.

TABLE 4   Comparison of Methods of Collecting Airborne Antigen Using Egg Albumin Aerosols in a Wind Tunnel and an ELISA Detection System

Sampler Type	Mean Recovery[a] (μg/l Air)	SE$_M$
High-volume filtration sampler	1.47	0.49
Large-volume electrostatic air sampler	2.12	0.65
Porton-raised impinger	0.24	0.06
Multistage liquid impinger	0.33	0.08
Sierra-Marple personal cascade impactor	1.26	0.38
Personal monitor filters		
Polypropylene	0.39	0.24
Glass fiber	0.42	0.06
Cellulose ester	0.59	0.21
Polycarbonate	0.87	0.10

[a]   All results are the means of three tests.
Note: SE$_M$, standard error of the mean.

VI. CONCLUSION

Determination of the causes of outbreaks of occupational asthma requires the detection of relevant allergens in the environment and identification of their sources. Microorganisms can be detected by culture using a range of samplers and the presence of allergens inferred. The detection of allergens present on nonculturable particles is more difficult. Immunoassay of samples collected on filters, using RAST inhibition assay, has allowed the presence of airborne allergens to be linked to specific IgE antibodies in the sera of asthmatic patients. However, the method is relatively insensitive and often requires large air samples. With some protein antigens, long sample times can result in poor recovery, possibly due to antigen binding to the glass fiber matrix or denaturation. Better recoveries may be obtained by sampling into liquid, using large-volume electrostatic air samplers, and increasing sensitivity of detection by using ELISA assays. However, there is still a need to improve assay sensitivity so that personal exposure can be determined using aerosol monitors or spill-proof microimpingers.

REFERENCES

1. Lacey, J. and Crook, B., Fungal and actinomycete spores as pollutants of the workplace and occupational allergens, *Ann. Occup. Hyg.*, 32, 515, 1988.
2. Lacey, J. and Lacey, M.E., Spore concentrations in the air of farm buildings, *Trans. Br. Mycol. Soc.*, 47, 547, 1964.
3. Pepys, J., Jenkins, P.A., Festenstein, G.N., Gregory, P.H., Lacey, M.E., and Skinner, F.A., Farmer's lung: thermophilic actinomycetes as a source of farmer's lung hay antigen, *Lancet*, 2, 607, 1963.

4. Lacey, J., *Thermoactinomyces sacchari* sp. nov., a thermophilic actinomycete causing bagassosis, *J. Gen. Microbiol.*, 66, 327, 1971.

5. Avila, R. and Lacey, J., The role of *Penicillium frequentans* in suberosis (Respiratory disease in the cork industry), *Clin. Allergy*, 4, 109, 1974.

6. Flindt, M.L.H., Pulmonary disease due to inhalation of derivatives of *Bacillus subtilis* containing proteolytic enzyme, *Lancet*, 1, 1177, 1969.

7. Blainey, A.D., Topping, M.D., Ollier, S., and Davies, R.J., Respiratory symptoms in arable farm workers: role of storage mites, *Thorax*, 43, 697, 1988.

8. Tee, R.D., Gordon, D.J., Lacey, J., Nunn, A.J., Brown, M., and Newman Taylor, A.J., Occupational allergy to the common house fly (*Musca domestica*): use of immunological response to identify atmospheric allergen, *J. Allergy Clin. Immunol.*, 76, 826, 1985.

9. Tee, R.D., Gordon, D.J., Hawkins, E.R., Nunn, A.J., Lacey, J., Venables, K.M., Cooter, R.J., McCaffery, A.R., and Newman Taylor, A.J., Occupational allergy to locust: an investigation of the source of the allergen, *J. Allergy Clin. Immunol.*, 81, 517, 1988.

10. Forster, H.W., Crook, B., Platts, B., Lacey, J., and Topping, M.D., Investigation of aerosols generated during sugar beet slicing, *Am. Ind. Hyg. Ass. J.*, 50, 44, 1989.

11. Griffin, P., Crook, B., Lacey, J., and Topping, M.D., Airborne scampi allergen and scampi peeler's asthma, in *Aerosols: Their Generation, Behaviour and Application, Aerosol Society Second Conference*, Griffiths, W.D., Ed., The Aerosol Society, London, 1988, 347.

12. Crook, B., Griffin, P., Topping, M.D., and Lacey, J., An appraisal of methods of sampling aerosols implicated as causes of work-related symptoms, in *Aerosols: Their Generation, Behaviour and Applications, Aerosol Society Second Conference*, Griffiths, W.D., Ed., Aerosol Society, London, 1988, 327.

13. Crook, B. and Lacey, J., Enumeration of airborne microorganisms in work environments, *Environ. Technol. Lett.*, 9, 515, 1988

14. Crook, B. and Lacey, J. Methods for enumerating airborne microorganisms in the work place, in *Airborne Deteriogens and Pathogens and Their Control*, Biodeterioration Society Occasional Publication No. 6, Flannigan, B., Ed., The Biodeterioration Society, Kew, U.K., 1989, 1.

15. Gregory, P.H., *Microbiology of the Atmosphere*, 2nd ed., Leonard Hill, Aylesbury, U.K., 1973.

16. Andersen, A.A. New sampler for the collection, sizing and enumeration of viable airborne particles, *J. Bacteriol.*, 76, 471, 1958.

17. Blomquist, G., Ström, G., and Stromquist, L.-H., Sampling of high concentrations of airborne fungi, *Scand. J. Work Environ. Health*, 10, 109, 1984.

18. Blomquist, G., Palmgren, U., and Ström, G., Improved techniques for sampling airborne fungal spores in highly contaminated environments, *Scand. J. Work Environ. Health*, 10, 253, 1984.

19. Popendorf, W., Report on agents, *Am. J. Ind. Med.*, 10, 251, 1986.

20. Hirst, J.M., An automatic volumetric spore trap, *Ann. Appl. Biol.*, 39, 257, 1952.

21. May, K.R., The cascade impactor: an instrument for sampling coarse aerosols, *J. Sci. Instrum.*, 22, 187, 1945.

22. May, K.R., Pomeroy, N.P., and Hibbs, S., Sampling techniques for large windborne particles, *J. Aerosol Sci.*, 7, 53, 1976.

23. Magill, P.L., Lumpkins, E.A., and Arveson, J.S., A system for appraising airborne populations of pollens and spores, *Am. Ind. Hyg. Ass. J.*, 29, 293, 1968.

24. Macher, J.M. and First, M.W., Reuter centrifugal air sampler: measurement of effective airflow rate and collection efficiency, *Appl. Environ. Microbiol.*, 45, 1960, 1983.

25. May, K.R., Multistage liquid impinger, *Bacteriol. Rev.*, 30, 559, 1966.

26. Eduard, W., Lacey, J., Karlsson, K., Palmgren, U., Ström, G., and Blomquist, G., Evaluation of methods for enumerating microorganisms in filter samples from highly contaminated occupational environments, *Am. Ind. Hyg. Ass. J.*, 51, 427, 1990.

27. Palmgren, U., Ström, G., Blomquist, G., and Malmberg, P., Collection of airborne micro-organisms on Nuclepore filters, estimation and analysis — CAMNEA method, *J. Appl. Bacteriol,* 61, 401, 1986.

28. Agarwal, M.K., Yunginger, J.W., Swanson, M.C., and Reed, C.E., An immunochemical method to measure atmospheric allergens, *J. Allergy Clin. Immunol.,* 68, 194, 1981.

29. Agarwal, M.K., Swanson, M.C., Reed, C.E., and Yunginger, J.W., Immunochemical quantitation of airborne short ragweed, *Alternaria,* antigen E and Alt-1 allergens: a two year prospective study, *J. Allergy Clin. Immunol.,* 72, 40, 1983.

30. Edwards, R.G., Beeson, M.F., and Dewdney, J.M., Laboratory animal allergy; the measurement of airborne urinary allergens and the effects of different environmental conditions, *Lab. Anim.,* 17, 235, 1983.

31. Price, J.A., Pollock, I., Little, S.A., Longbottom, J.L., and Warner, J.O., Measurement of airborne mite antigen in homes of asthmatic children, *Lancet,* 336, 895, 1990.

32. Swanson, M.C., Agarwal, M.K., and Reed, C.E., An immunochemical approach to indoor aeroallergen quantitation with a new volumetric air sampler: studies with mite, roach, cat, mouse and guinea pig antigens, *J. Allergy Clin. Immunol.,* 76, 724, 1985.